# Indigenous Agency in the Amazon

# Indigenous Agency in the Amazon

## *The Mojos in Liberal and Rubber-Boom Bolivia, 1842–1932*

Gary Van Valen

The University of Arizona Press
Tucson

THE UNIVERSITY OF ARIZONA PRESS


www.uapress.arizona.edu

Library of Congress Cataloging-in-Publication Data

Van Valen, Gary.
Indigenous agency in the Amazon : the Mojos in liberal and rubber-boom Bolivia, 1842–1932 / Gary Van Valen.
p. cm.
Includes bibliographical references and index.
ISBN 978-0-8165-2118-0 (cloth : alk. paper) 1. Mojo Indians—Bolivia—Beni—History. 2. Rubber industry and trade—Social aspects—Bolivia—Beni. 3. Millennialism—Bolivia—Beni—History. 4. Agent (Philosophy). I. Title.
F3320.2.M55V36 2013
678′.2098442—dc23 2012033141

Publication of this book is made possible in part by the proceeds of a permanent endowment created with the assistance of a Challenge Grant from the National Endowment for the Humanities, a federal agency.

Manufactured in the United States of America on acid-free, archival-quality paper containing a minimum of 30% post-consumer waste and processed chlorine free.

18 17 16 15 14 13 6 5 4 3 2 1

# *Contents*

# *Illustrations*

## Figures

## Maps



## Tables

# *Acknowledgments*

This book would not have been possible without the help of many people. First of all, I would like to thank my parents, John and Emma Van Valen, for always being there for me. Their love, financial and emotional support, and encouragement to follow my dreams made this book and many other good things in my life possible. I also would not have chosen the path that led to this book without the model of intellectual curiosity that my brother Scott has always provided. To my wife Veruska, whom I met during the course of my research, I wish to express my gratitude for her love and support. I thank my sons John P. and Joseph for an understanding beyond their years that papá cannot spend all day playing. Although none of these family members are particularly interested in history, they have all demonstrated a remarkable patience with the multiyear project that this book represents.

I deeply appreciate the financial support given by the University of New Mexico's Latin American and Iberian Institute in the form of two years of Dissertation Fellowships and two Research Travel Grants that helped to fund the first two of my five trips to Bolivia. I am also grateful to Roanoke College for helping fund my final research trip through a Starter Grant. I would like to thank the Marion, Petersen, and Avila families of Sucre for their friendship, to which I owe much of my understanding of both Latin American culture and my fluency in the Spanish language. Alberto Marion Argandoña and Richard Petersen Kelly proved to be especially good friends, and I am thankful for their help. I would also like to thank my friend and colleague Javier Marion for providing me contact with his family and for his friendship in general.

In Sucre I owe thanks to three directors of the Archivo y Biblioteca Nacionales de Bolivia, René Arze Aguirre, Hugo Poppe Entrambasaguas, and especially Marcela Inch Calvimonte, and to staff members including Lidia Ortega, Leonor Ferrufino, and María Eugenia Peñaranda. I would like to thank the personnel of two other collections in Sucre as well. The Biblioteca de la Casa de la Libertad provided me access to an issue of the newspaper *Los Debates* that was missing from the collection of the Biblioteca Nacional. Without Miguel Antelo's report in that issue, I would have been unable to write chapter 4 of this book in its present form. Documents from the Centro Bibliográfico Documental Histórico Chuquisaca also enhanced my study. In Tarata, Fray Mauricio Valcanover at the Colegio de San José was extremely helpful in locating and photocopying relevant Franciscan sources dealing with the mission of San Lorenzo. For Santa Cruz, I would like to thank the Archbishopric of Santa Cruz for allowing me access to documents from their archive, and to archivist Luis Enrrique Rivero for locating Bishop Prado's report there. Rosario Vargas and Paula Peña also opened the archive of the Universidad Autónoma Gabriel René Moreno to me, with fruitful results. I am also grateful to Roberto Escóbar for his hospitality and general help in Santa Cruz.

I would like to express my gratitude to my dissertation committee at the University of New Mexico for all of their time and effort. My chair, Kimberly Gauderman, was especially devoted to helping me improve my writing and analysis as well as aiding my job search. I can only say that after having gone through the first years of teaching at two successive institutions, I appreciate her taking on the duty of heading my dissertation committee during her first year even more. Linda Hall and Judy Bieber both contributed to several semesters of my classroom education before providing valuable commentary on my dissertation. Linda has continued to help in numerous ways, and Judy deserves credit for suggesting that agency might better describe Mojo actions than resistance. I would also like to thank the outside reader of my dissertation, Suzanne Oakdale, for her time and comments. I consider all of my advisors friends and role models.

I also benefited from the advice of the participants in the fall 2002 dissertation seminar and thank them for reading and commenting on chapter 5. I am grateful to have known Robert Himmerich y Valencia and the late Robert Kern, who were also teachers, friends, and role models during my time at UNM. I thank the numerous friends and colleagues who in one way or another helped me make it through my PhD program, including Jonathan Ablard, Stan Hordes, Nancy Brown-Martínez, Andrae

Marak, James Martin, and Kim Suina, among others. To James I am also grateful for his fine work in preparing the two maps in this book.

New friends and colleagues have helped me carry this project beyond the dissertation stage. The companionship and ideas of two remarkable scholars of the Bolivian missions, María José Diez and Manuel Plaza, made my final research trip especially enjoyable. My friends and colleagues Aran MacKinnon and Chuck Lipp took the time to comment on my new introduction. Alcira Dueñas organized a session at the American Historical Association meeting of 2008 so that we could present our work together, and Joanne Rappaport gave valuable commentary. I am also grateful for the friendship and support of West Georgia colleagues including Jonathan Goldstein, Steve and Martha Goodson, Euisuk Kim, Elaine MacKinnon, Sal Peralta, Sandy Pollard, Nadya Popov, Colleen Vasconcellos, and Dan Williams.

I must acknowledge my debt to other scholars who have previously written on the Mojos and the history of Beni: Jürgen Riester, David Block, Zulema Lehm Ardaya, James Jones, Jorge Cortés Rodríguez, Manuel Limpias Saucedo, Rodolfo Pinto Parada, Arnaldo Lijerón Casanovas, and Enrique Jordá Arias, among others. Without the framework provided by their writings, I would have been unable to make much sense of the history of the Mojos or of Beni. David Block's *Mission Culture on the Upper Amazon* inspired me to investigate a later period of Mojo history, and I am grateful for the interest and support he has shown for this project and for my career. The library staffs of the universities of South Carolina, New Mexico, West Georgia, and of Roanoke College also helped make this study possible, either through their own collections (especially that of New Mexico) or their interlibrary loan services. Cláudio Barbosa and Aline Siqueira of the Museu de Arte Moderna in Rio de Janeiro made possible the use of Frisch's 1865 photograph of Mojos in Manaus, an image that expresses the strength of these people and inspired me as I labored to convey a similar impression through my writing. I also would like to thank University of Arizona editors Patti Hartmann for taking on this project and Kristen Buckles for seeing it through, the anonymous reviewers for their time and valuable comments, and all others who helped in the final preparation of this book.

Finally, to the Mojos themselves, I would like to say that I have gained a tremendous amount of respect for their ability to survive no matter what the circumstances. I somewhat regret that my research methodology required that virtually all my time be spent in archives far from their homeland. Although indigenous historians are appearing increasingly in Bolivia, they will probably not have the resources to spend years in the archives for the foreseeable future. I have become increasingly aware of the oddness of a North American scholar being able to build a career researching Latin American indigenous people who do not have the luxury of conducting their own archival research, a situation that can only be explained by the division of the world into rich and poor countries. I consider this book the first step in making my findings accessible to the modern Mojos and hope that they will be able to draw strength from an increased knowledge of the struggles of their ancestors.

# Indigenous Agency in the Amazon

# *Introduction*

On a hot day in May 1887, a group of Bolivian soldiers halted their march across a flat savanna laced by lakes and tributaries of the Amazon. They were taking ten indigenous prisoners back to their headquarters in Trinidad, the local capital, but stopped to administer several hundred lashes to each. The soldiers took special delight in whipping an eighty-year-old man before finishing him off with gunshots. They saw that old man, Andrés Guayocho, not as the pathetic figure who begged them for a drink of water before dying, but as the leader of a major indigenous rebellion. Guayocho was a Christian shaman who used his ability to throw his voice to convince the Mojo Indians that Jesus Christ and deceased Mojo leaders were speaking to them, and attracted hundreds of Mojos away from the ex-Jesuit mission of Trinidad to a new settlement in the backlands. The local government ruthlessly crushed Guayocho's millenarian movement not so much because of its religious unorthodoxy, but because it deprived the white elite of the Mojo labor necessary for the extraction and export of rubber for the world market.

The Mojos who survived this massacre shared the same heritage that had informed Guayocho's movement: a mixture of indigenous culture and the Catholic, colonial culture introduced by Jesuit missionaries in the late seventeenth century. However, those who began returning to Trinidad in 1888 asked that they be respected not just as natives with original rights to territory or as fellow Christians, but also as fellow citizens with constitutional rights. In the ensuing years, these same Mojos served in the local military, celebrated Bolivian Independence Day, sang the Bolivian national anthem in their church choirs, sought out modern educations, and participated in private land transactions. They clearly knew about and used the European ideology of liberalism, with its language of citizenship and equality before the law. The contrast between millenarian rebellion and participatory citizenship could not be greater, yet it gives some indication of the incredible range of strategies employed by the Mojos in their relations with the state and the local white elite.

During the nineteenth century, the Mojos brought together indigenous culture, mission Catholicism, and liberalism to form a hybrid identity, which they used to defend their interests. The Mojos' hybrid identity gave them an expanded capacity for what has been called "agency." Agency, the freedom to act, can vary among individuals, groups, and classes in any given historical situation. It is relational, for people operate within a web of social relationships, pushing or pulling against other individuals or groups with their own potential for agency. Any attempt to renegotiate power relations is agency, from engagement in overt resistance such as a rebellion or millenarian movement, to the use of discourse, or language encoding a vision or plan for social relations. The value of agency as a tool of historical analysis lies not in insisting that all people have unfettered agency at all times, but rather in studying what factors promote or limit agency in certain contexts. Agency is especially useful for the analysis of complex indigenous societies like the Mojos, whose internal differentiations can lead individuals or subgroups to take different actions based on what will benefit each the most. It thus allows for a more flexible treatment of indigenous history than previous theoretical frameworks permitted, whether ethnographic ideas of closed communities from the 1930s through the 1950s, Marxian accounts of peasant rebellion popular in the 1960s and 1970s, or theories of resistance employed in the 1980s and 1990s.[1]

In this book, I weave the theme of agency through an ethnohistory of the Mojos during the liberal reforms and the rubber boom. Until now, this period in the Mojos' history has largely been an untold narrative. Their history, exceptional because of the Mojos' unique historical and geographical situation, has at the same time a much wider significance. Our incomplete understanding of the varieties of Latin American liberalism and of the rubber boom is further illuminated by a detailed examination of Mojo history in the late nineteenth and early twentieth centuries.

The challenges faced by the Mojos reflect common themes in the historiography of nineteenth-century Latin America, when newly independent countries introduced liberal political and economic policies and expanded commerce (including the export of raw materials) with the outside world.

Most scholars dealing with this era have interpreted the period of liberalism and export economies as a time of suffering for poor and indigenous people.[2] But by looking through the prism of agency, we can see that the Mojos were not inherently victims or powerless in the face of changes imposed by others, and that they would prove to be active participants in shaping the course liberalism and the rubber boom would take in their region.

Liberalism, a compilation of ideas from the European Enlightenment, came to inform political thought throughout the Americas by the mid-nineteenth century. Liberal thinkers discarded inherited, preconceived notions of state and economy, including the legal inequality and state-licensed monopolies of the eighteenth century. They conceived of human reason as the key to creating new and better societies, and assumed that reason dictated a new approach to government's relationship with individuals and the economy. In an ideal liberal society, corporate groups with unique and differing legal privileges would be replaced by a nation of individuals given equal treatment under the law, and individuals would be citizens, not subjects, of the state. The ideal liberal economy would be of the "free market" type, free from state intervention, ruled by the law of supply and demand. Liberals came to believe that by liberating the individual and the economy, they could open the doors to progress, and that education would ensure the continuing progress of the nation.

This summary of liberal ideals should not be taken as proof that liberalism was ever just one thing. As a socially constructed language on human political and economic relations, enacted in an arena of unequal power relations, liberalism was a discourse and not just an ideology. Liberal thought was constantly evolving, and every liberal thinker or government adapted liberal ideas to specific conditions of time and place. Local debate and interpretation determined how far the reorganization of society went, who was allowed to vote, how much government disentangled itself from the economy, and how much it promoted new forms of education. Interpretations of liberalism also changed over time, as in Europe, where liberals' priorities evolved as they weighed their desire to replace absolutist government against their fear of mob rule and another Reign of Terror. Thus the leaders of newly independent Latin American countries tended to adopt a modified liberalism that emphasized limiting executive power through constitutions and effective legislatures; providing voting rights to some, but not all, men; declaring equal treatment under the law to all, regardless of capacity to vote; rationalizing and homogenizing administration; and lifting old restrictions on manufacturing and commerce. Even these ideas were contested, and yet no matter how they adapted the basic tenets, liberals always felt their beliefs to be reasonable, and as "self-evident" and dependable as other laws of nature revealed by scientific inquiry.[3]

Many Bolivian elites were attracted to this new vision of society. To them, liberalism seemed rational and progressive, and offered the prospect of modernizing and increasing material prosperity without a complete overthrow of social hierarchy. But in Bolivia as in Europe, liberalism was not one thing, but an interpolation of new ideas into existing local societal forms and patterns of power distribution. Governmental authorities and dominant groups put it into practice in ways that suited their local needs, which in turn influenced its language. Thus Bolivian elites often continued an older, colonial discourse on Indians behind the new language of equality and citizenship. This colonial way of ordering society treated Indians as separate and by nature inferior due to the Indians' imagined subjection through conquest, their supposedly inadequate reasoning ability, and their perceived miserable, impoverished condition. The national government also continued colonial policies of protecting Indians in return for tribute, unpaid labor, and adherence to Catholicism. While creole elites juggled and combined colonial and liberal discourses, Bolivian indigenous peoples like the Mojos found themselves confronted with the new ideas and created their own interpretation of liberalism. Coming to know something of the language of liberalism, they used it to influence how other societal actors treated them. Mojo history provides a concrete example of how the state, the local creole elite, and indigenous people negotiated the meaning of liberalism on the ground.

The few examinations of the use of liberalism by Indians for their own benefit have only appeared in recent years and concentrate on Mesoamerica and the Andean highlands. In a significant change in interpretations of the liberal era, researchers have discovered that Indians did not simply cling to the past and blindly resist all of the liberal reforms. Indian leaders, in particular, were willing to actively participate in liberalism, and to demand the rights of citizenship and public education it promised, as a means to preserve their communities.[4] I find that traditional interpretations of the impact of liberalism on indigenous people do not fit the historical experience of the Bolivian Amazon in this era. Even though the white elite soon lost interest in the political principles of an inclusionary, idealistic liberalism, the Mojos, especially their elite, learned to use these ideas to increase indigenous agency.

Just as independence created new possibilities for Latin Americans to experiment with liberalism, it also opened Latin America to a world market dominated by the industrialized nations of Europe and North America. Their industries and industrial societies consumed vast quantities of raw materials, and Latin America helped meet the demand with its resources. In the Amazon basin during the years considered by this book, local elites and foreign capitalists extracted rubber for export in response to this external demand. According to the traditional historiography, the development of Amazonian rubber extraction resulted in massive movements of people into and within the rubber zone, the conversion of rubber workers into debt peons, and the decimation of indigenous communities through exposure to disease, warfare, and virtual enslavement. Although this depiction largely holds true for indigenous groups living outside state authority in northern Bolivia, it cannot be applied to the Mojo region, which was transformed without being devastated by the rubber boom. The Mojos themselves were changed by their experience of the rubber boom, but retained their sense of community. By adapting to political and economic change, they remained Mojos.

The Mojos were not destroyed by the rubber boom, and during some five decades of Bolivian participation in the export of rubber, their opportunities for independent action varied but remained quite high. The period from 1860 to 1880 was a time of increased agency as the Mojos and other ex-mission Indians found new opportunities beyond their home towns. The Mojos' ability to act was severely challenged only during the decade of 1880–90, and Guayocho's movement emerged in response. The years from 1890 to 1910 can best be characterized as a time of increasingly creative responses to a narrowing space for action. The greatest limits on indigenous agency came only after 1910, when the rubber boom ended, indigenous participation in the export economy became limited, and cattle ranching replaced rubber as the principle economic activity of the Bolivian Amazon. After a thorough examination of Mojo history, the idea that the rubber boom was prejudicial to all Indians at all times becomes untenable.

The history of the Mojos offers material for a fruitful comparison with the experience of other indigenous groups in the era of liberalism and export economies. The history of frontier, ex-mission Indians in Latin America is a relatively recent and still-developing field, and the geographic and environmental uniqueness of the Llanos de Mojos, the frequently flooded plains of eastern Bolivia, makes the Mojo experience especially worthy of study.[5] Recent scholarship on frontiers and borderlands highlights cultural contact, social fluidity and mobility, and the ability to negotiate identity, all of which can be found in the Bolivian Amazon.[6] Because the Bolivian state regarded its Amazon region as a frontier, another discourse, that of civilization and barbarism, figured prominently in negotiations between the ex-mission Indians, the state, and the local white elite. Independent Indians, usually and disparagingly called *indios bárbaros* (barbarous Indians), also played a part in such negotiations, and ex-mission Indians like the Mojos were able to use both the image and reality of such "bárbaros" in surprisingly innovative ways to expand their own agency. As the center of their world, the Bolivian Amazon was not necessarily a "frontier" to the Mojos, but they understood and used the state's conception of their homeland for their own purposes.

The former mission Indians were uniquely well equipped to contest the projects of political and economic elites precisely because of their dual heritage as members of states and residents of the Amazon frontier. Former mission Indians like the Mojos could exercise agency in a great many forms, including migration, millenarianism, participation in the politics and discourses of the dominant society, and manipulation of the categories of "civilized" and "savage" used in one of those discourses. The Mojos exploited their position on the Amazon frontier, with its weakness of political authority, low population density, fluidity of identities, and proximity to independent Indian groups. They took advantage of opportunities that the Amazonian environment provided to move their cattle herds and slash-and-burn agricultural plots away from areas controlled by exploitative elites. They also capitalized on their long history of incorporation in wider political and economic systems, which made them particularly adept at exercising agency through participation and manipulation of categories, and provided them with leaders skilled in dealing with white authorities.

Earlier interpretations of modern Latin American indigenous peoples tended to limit their survival strategies to sealing off communities from the outside world or engaging in fruitless rebellions against outside oppressors. I find that the Mojos never tried to isolate themselves, and they never completely rejected change. Nor were they just hapless victims of the rubber boom. While Guayocho's millenarian movement may appear at first to be a singular and failed resistance to profound change, it is better understood as one facet of native agency. Guayocho's movement thus was not a one-time reaction to an external stimulus, but an episode in a long history of Mojo action. And as we shall see, his movement was not a failure, but rather the tumultuous birth of a new Mojo community that survived and engaged effectively with the outside world for another three decades.

The chapters that follow introduce the reader to the Mojos and their homeland, and then delineate their many attempts to renegotiate power relations by making dynamic and well-thought-out choices for action.

Chapter 1 presents a geographical, ethnological, and historical background of the Llanos de Mojos up to 1842. Chapter 2 chronicles the liberal reforms that created the Department of Beni and dismantled much of the mission organization, and assesses their impact on the Mojos. Chapter 3 discusses the development of the Bolivian rubber boom and examines the varied forms of agency used by the former mission Indians from the 1860s to the 1880s. Chapter 4 deals with the millenarian movement among the Mojos in 1886–87 as one response to the rubber boom. The next two chapters trace the histories of two groups of Mojos from the suppression of the millenarian movement until 1932, just before the Chaco War brought sweeping change to all sectors of Bolivian society. Chapter 5 recounts how the Mojos of the millenarian settlement of San Lorenzo maintained a good deal of control over their fate from the 1890s to the 1920s. Chapter 6 follows developments among the Mojos of the former mission towns of Trinidad and San Ignacio in the same period. The conclusion reiterates some of this study's important discoveries and relates them to the broader themes described in the introduction. Throughout the book, the reader will discover the unique and inspiring story of a small group of indigenous people who, through their own resourcefulness at combining tradition and change, secured their group's cultural survival amidst sweeping political and economic forces of global scale.

CHAPTER ONE

# *The Llanos de Mojos*

## Geographical Situation

Few countries are as obviously divided into two parts by their physical geography as is Bolivia. When outsiders consider Bolivia, they usually think of the Andes Mountains, llamas, potatoes, and Indian peoples descended from the Incas and earlier civilizations. This description fits the western portion of the country, the *tierras altas*, comprised of a cold, high plateau (the *altiplano*) and warmer, lower valleys, all over two thousand meters in elevation. Though it comprises less than a third of the modern nation's area, the highland region of Bolivia has always hosted the densest populations, the most complex cultures, and the centers of political power. This story takes place in the other part of Bolivia, known as the *tierras bajas* for its low elevation (below two thousand meters), or as the *Oriente* for its location in the eastern part of the country. The modern Oriente is comprised of three major political divisions, the departments of Santa Cruz, Beni, and Pando, along with parts of other, highland-centered departments. This study deals primarily with events in the Department of Beni from 1842 to 1932.

Beni lies within the basin of the Amazon River and is traversed by northward-flowing rivers that, while enormous in themselves, are but minor tributaries of the Amazon. The Beni River flows along the western boundary of the department, the Mamoré River through the center, and the Iténez River (known as the Guaporé in Brazil) along the boundary between Beni and Brazil. The Iténez flows into the Mamoré in the northern part of the department (Brazilians would say that the Mamoré flows into the Guaporé). The lower Mamoré (or lower Guaporé) joins the Beni

*Figure 1.1.* Beni River at Cachuela Esperanza. Author photograph.

River at the northern tip of the department to form the Madeira River, a major tributary of the Amazon. Many smaller rivers cross the territory of the department and flow into the Beni, Mamoré, or Iténez systems.

The Department of Beni is almost all a plain, known as the Llanos de Mojos or Llanos de Moxos, which is nowhere higher than five hundred meters in elevation. Climatically, Beni forms part of what Spanish Americans traditionally called the *tierra caliente* (hot land); a more modern definition would be those lands with a tropical rainy climate (type A climates in the Köppen system). While areas farther north in Amazonia, including the Department of Pando, have heavy rainfall all year that supports tropical rain forest, Beni has a long dry season which lasts from May to September. Because of the dry season, tropical forest can only be found on Beni's far northern edges and along parts of the major rivers. The rest is tropical savanna: grasslands dotted with palm trees of different types and occasional patches of forest. There are no seasons based on temperature change as there are on rainfall. Since Beni lies only ten to seventeen degrees south of the equator, with no mountains, the climate is hot year round. The only exception is when a cold *surazo* (south wind) blows in from Argentina, generally between May and July, and then the temperature can lower dramatically.[1]

The numerous rivers, lakes, and arroyos of Beni make most parts of the department easily accessible by water. Communication with the Atlantic

Ocean via the Amazon River would be easy as well, but for the presence of eighteen major rapids and waterfalls on a 380-kilometer stretch of the Mamoré and Madeira rivers where they flow over the hard rocks of the Brazilian Shield.[2] These rapids, or *cachuelas*, also slow the flow of water so much that the low-lying, upriver Llanos de Mojos are usually flooded each rainy season. Some years experience worse floods than usual; in the period under consideration, there were great inundations in 1853, 1865, 1872, 1873, 1886, 1895, 1896, 1907, 1921, 1928, 1929, and 1931.[3] Recent core sampling of sediments in the Llanos de Mojos has produced evidence that the largest floods occur during the so-called La Niña years of the El Niño–Southern Oscillation, when cold ocean temperatures generate higher rainfall in the central Andes.[4]

The only landforms that escape flooding are relatively small, slightly elevated parcels of land called *islas* (islands) or *alturas* (high grounds).[5] When the floods come, every human and every land animal, wild or domestic, crowds onto the islas.[6] These high grounds became even more valuable after the introduction of cattle in the late seventeenth century, and since then they

*Map 1.1.* The Llanos de Mojos. Map by James Martin.

have acquired the additional name of *rodeos de ganado*, or cattle corrals.[7] The Llanos de Mojos are thus the opposite of dry lands such as northern Mexico and the western United States, where permanent water sources are rare, valuable, and necessary for the exploitation of surrounding lands. In Beni, water is abundant, and the most valuable lands are those that stay dry.

## Native Prehistory

Human beings have inhabited the Llanos de Mojos for several thousand years, but we know little of what happened in those millennia before Spanish documentation began. The region's extreme linguistic diversity suggests several different migrations into the area, though Swedish anthropologist Alf Hornborg has cautioned against a "billiard-ball model of migrating, essentialized 'peoples,'" and advanced the thesis that the Arawakan languages spread through river trade rather than movement of whole ethnic groups.[8] The river systems of the Llanos de Mojos certainly facilitated the spread of languages, whether through migration, trade, or both.

Almost all of the llanos peoples shared elements of what has been called Tropical Forest Culture, including pottery, hammocks for sleeping, effective river craft, and agriculture based on tropical root crops, especially manioc.[9] Only the Sirionó, a hunter-gatherer people who lived east of the Mamoré, had no canoes and almost no agriculture. Organized into bands, they had the least complex society in the region, moving from camp to camp, building and abandoning their communal houses. Because their hunting parties raided Spanish and Bolivian boats and settlements in order to steal iron tools, they long remained indios bárbaros in the eyes of the Hispanized population.[10]

The most complex cultures lived along the Mamoré and Iténez river systems, and included the Mojo, Canichana, Movima, Cayubaba, Itonama, and Baure peoples. At Spanish contact, the Indians we refer to as "Mojos" were comprised of at least twenty small ethnic groups (called *parcialidades* in Spanish) whose people spoke related Arawakan dialects. None called themselves Mojos; the term originated from the Spanish expectation to find a rich, Eldoradolike society in the region ruled by the "Gran Moxo," a name of uncertain origin. This and other legends of rich kingdoms in the region stemmed from accounts of the Inca Empire told to the Spaniards by the Guaraní Indians in Paraguay.[11]

Most of these people belonged to nonranked kinship-based societies with no authority above the individual community, or what many anthropologists would call tribes. Some scholars have suggested that the Arawakan-speaking Mojo and Baure developed chiefdoms, ranked kinship-based societies where the chief and his kin group have authority over other kin groups, and sometimes over several communities. There certainly seems to be archaeological evidence of coordination in labor above the level of the single community, as the Llanos de Mojos are covered with the remains of ancient earthworks. From at least the ninth century BC (according to recent radiocarbon dating), the llanos natives piled earth into raised fields, mounds, and causeways connecting different islas, all in order to escape the yearly floods or to increase production of agriculture or fish. Even so, other scholars doubt the existence of any authority above the individual village because of a lack of ethnohistorical evidence.[12]

The contact-period Mojo chief, or *achiaco*, had no permanent coercive authority and exercised leadership mostly during war or hunting expeditions. The division of labor was determined by age and sex, not by any class system (though anthropologist Alfred Métraux argued that war captives formed an "incipient servile class"). Kinship was one of the principal ways of integrating society. The pre-mission Mojos had extended families that lived in houses and kitchens ranged around a village plaza. There is linguistic and historical evidence that extended families existed, but scholars have offered differing views of how many family members actually lived together in one house. At any rate, the very large communal house, or *maloca*, common among many Amazonian peoples, did not exist among the Mojos.[13]

Ritual was another way of integrating society. On one side of the plaza stood what the Spanish called the *bebedero*, or drinking house, the center of Mojo ritual life. Mojo men and sometimes women used the bebedero for drinking bouts with *chicha* (beer, most often made from corn). They would meet there for ten to twelve feasts a year, and for events such as going off to war, on a long voyage, or to hunt a jaguar. The Mojos kept skulls of both jaguars and human enemies in the bebedero, and some large villages had more than one such building. Like most Native Americans that used alcohol prior to European contact, the Mojos' drinking was invested with religious significance. Fasting and dancing were other important expressions of Mojo spirituality.[14]

Mojo ritual specialists performed the priestly function of offering chicha to the gods in the bebedero. The same individuals performed the shamanic function of communicating with spirits for purposes of curing and divination. In order to contact spirits, they sometimes drank a drug made from a plant called *marari*. The general term for a shaman in Mojo, *tiarauqui*, means "one who has clear eyes," and shamans acquired their clear vision by having the powerful juice of vines or herbs squeezed into their eyes. Like other shamans the world over, they assumed their role after some traumatic experience. Mojo men and women became tiarauquis after recovering from an illness, encountering a spirit, or surviving the attack of a jaguar or caiman (considered powerful spirits in Mojo belief, with jaguars given greater importance in Spanish sources). Those shamans wounded by jaguars acquired the power to communicate with jaguar spirits and were called *comocoes*.[15]

The belief system of the Mojos tied them closely to the territory they inhabited. Each parcialidad had its own god, which could not abandon a specific district, while other spirits inhabited particular hills or lakes. The Mojos believed that their ancestors had come from a place within their native territory and also that they could communicate with their ancestors. A great many other spirits pervaded nature, forming two groups associated with the forest and the water. The jaguar and caiman, as the dominant predators in these respective domains, were thus among the most important powers in Mojo belief.[16]

## The Arrival of the Spanish

The Spanish first entered the Llanos de Mojos in the sixteenth century, and many subsequent expeditions explored the region in that and the following century. They generally arrived from the south, from the settlement of Santa Cruz de la Sierra established by Paraguayan Spaniards and their Guaraní Indian allies in 1561, and transferred to its present site by 1621.[17] In 1682, after several preliminary visits and eight years of permanent residence among the Mojos, Jesuit missionaries began to congregate the Indians of the Llanos de Mojos into towns.[18] The llanos natives accepted the Jesuits because such an arrangement offered more benefits than drawbacks. The missionary presence afforded protection from slave raids from Santa Cruz in the seventeenth century and from Brazil in the eighteenth century. Alliance with the Jesuits also meant regular access to the two most important commodities that the llanos could not provide: salt and iron.

Christianity was clearly foreign to the llanos natives' worldview, but they and the Jesuits managed to find some religious common ground. When the fathers demonstrated an ability to cure dysentery and other illnesses, the natives could understand them as shamans.[19] Although the Jesuits destroyed the bebederos, the Catholicism of the missions allowed for the continued expression of spirituality through fasting and dancing, and chicha continued to be made and drunk.[20] Shamanism and belief in spirits somehow survived the Jesuit period as well, for they are noted at the end of that period and reappear in the nineteenth and twentieth centuries.[21]

Sustained contact brought the biological transfers known to historians as the Columbian Exchange to the Llanos de Mojos. The llanos natives benefited by acquiring a great many new plants and animals. The Spanish introduced Old World crops that prospered in many tierra caliente regions of their empire, including sugarcane, rice, citrus fruits, and plantains, as well as tropical crops like cacao from elsewhere in the Americas. The Jesuits also brought cattle to replace game animals, even though game hunting began to matter less as the gathering of the Mojos into villages restricted the time they had used for hunting. But, as in other areas of the Americas, the Llanos de Mojos suffered a massive decline of the native population after the introduction of Old World diseases by the Europeans. The resulting epidemics of course brought no benefits to the Indians, but they did contribute to a continued need for Jesuit protection by reducing their numbers.[22]

The Jesuits expanded from the lands of the Mojos to establish missions among the Canichanas, Movimas, Cayubabas, Itonamas, and Baures before their expulsion by the Spanish government in 1767. Although the mission population grew to about 30,000 people by 1720, it leveled off and then declined to 18,535 in 1767, because of epidemics and low fertility due to a shortage of women.[23] But, according to historian David Block, the Jesuits "left behind a resilient mission culture based on firm foundations." Block defines "mission culture" as "new systems—biological, technological, organizational, and theological" that allowed for the survival of indigenous elements in modes of subsistence and leadership, and thus represents an accommodation between native culture and elements contributed by the Jesuits. In sum, the mission experience created a Christian Indian population settled in towns with communal lands and cattle herds, some knowledge of the Spanish language, and an increasing resistance to Old World disease.[24]

## The Organization of Mission Society

The mission experience significantly reorganized the social structure, government, and productive activity of llanos societies. The Jesuits divided the Indians into two groups: an elite called the *familia* and the common people, or *pueblo*. Although the nineteenth-century French visitor Alcides d'Orbigny described them as "hereditary classes," these two groups were not classes solely based on economic benefit and exploitation, but were more like the orders or estates of early modern Europe, in that they were hierarchically ranked categories based on people's function in society. The familia was composed of the indigenous political leadership and those Indians trained as singers, religious assistants, cooks, carpenters, weavers, blacksmiths, and other types of artisans. Some of the chiefs and even some of the shamans of pre-mission society became part of the familia. The pueblo represented the labor force under direction of the familia; its members handled agricultural work, cattle herding, and transport, and their surpluses provisioned the familia and the Jesuits. Where several small ethnic groups had been resettled in one town, as at all of the Mojo settlements, members of the pueblo order often identified with specific parcialidades descended from the original ethnic groups. Extended families continued to exist and could include people from both the familia and the pueblo. Kinship thus provided a means of bridging the gap between the two orders and gave commoners access to elite patronage. Mission Indians lived in multifamily houses, and the residents of any one house most likely belonged to the same pueblo parcialidad or familia industry.[25]

In 1701, the Jesuits introduced new town governments similar to those they had created in Paraguay and based ultimately on those of Spanish municipalities. These governments, called *cabildos indigenales*, were staffed entirely by Indians of the familia order, and their principal purpose was to mediate between the Spanish authorities (in this case, the Jesuits) and the mass of indigenous people.[26] Like other Indian elites throughout Spanish America, the familia thus occupied the ambiguous and sometimes uncomfortable position between rulers and subjects. The cabildos organized the labor and the religious fiestas required by the Jesuit missions.[27]

Orbigny says that a mission-period cabildo was comprised of eight members, each of whom had a silver-tipped staff (*bastón*) as a symbol of authority. The eight included a *cacique*, an *alférez*, two *tenientes*, two *alcaldes de familia*, and two *alcaldes de pueblo*. Lesser officials, who each carried a black staff of office (*vara*), included a *mayordomo* and *segundo* for each industry of the familia, a *capitán* and segundo for each parcialidad of the pueblo, an *alcalde de estancia* who attended the cattle herds, and a *fiscal* who executed sentences.[28] It should be noted that both the creation of Indian cabildos and the investment of Indian officials with staffs of office was the normal Spanish policy throughout their American empire.

The cabildos orchestrated the material production of mission agriculture and industry, known as *temporalidades*, which included cacao, cotton thread, cotton cloth, sugar, cane alcohol, tallow, beeswax, furniture, musical instruments, and liturgical textiles. The cabildo turned these products over to the Jesuits, who distributed exotic products such as salt, woolen cloth, iron, knives, paper, wheat flour, and wine for mass in return. The Jesuits, the cabildo, the familia artisans, and the pueblo laborers all thus benefited to varying degrees from this redistribution of goods and labor. The redistribution of the temporalidades system is evidence that the Jesuits not only introduced local modifications in government, social organization, production, and subsistence, but also incorporated the Mojos and their neighbors into a wider Spanish American society. To fund the missions, the Jesuits relied on profits from sugar haciendas on the north coast of Peru, royal contributions, private donations, and the wealth generated by mission exports. They traded these products by routing them through Jesuit colleges throughout the central Andes, or else sold them directly in Spanish markets in Santa Cruz and Cochabamba.[29] As expert boatmen, the llanos natives provided the paddlers who transported this produce to those markets.

Whether members of the pueblo who had familia kinsmen in the cabildo were able to avoid certain labor obligations is unknown. In general, individuals could not hope to disobey the cabildo without punishment, and cabildo officers had the authority to administer whippings to the people of the mission. Orbigny observed that Indians were whipped merely for being distracted and not properly saluting the cacique. Whether or not whippings were used as punishment before the Spanish arrived, they were certainly in accord with Spanish colonial practice. Most Indians were commoners, and early modern Europeans believed that corporal punishment was the appropriate way to discipline commoners. The Spanish colonial discourse on the Indians also stressed their inadequate reasoning ability and equated them with dependent children, who would only be able to understand corporal punishment when they misbehaved.[30] That the Jesuits and cabildo officers shared these attitudes should be kept in mind to avoid a utopian conception of the Jesuit missions.

## The Former Mission Indians under Secular Spanish Rule

Immediately after expelling the Jesuits, the Spanish replaced them with secular "priests" (*curas*) who came from Santa Cruz and who were not necessarily ordained. The curas co-opted the temporalidades system for their own benefit, trading mission products to Santa Cruz, the highlands, and Brazil through relatives and friends for personal gain. The curas cared little about the former missions' economic health and slaughtered vast numbers of cattle to sell tallow. Spanish governors, representing royal interests, did little to restrain the curas until 1786, when Lázaro de Ribera took office. Ribera placed royal administrators in each former mission town and limited the curas' power to religious matters. He also rebuilt the cattle herds and greatly increased the production of cacao and cotton textiles in a successful plan to make Mojos profitable. Ribera's successors had to deal with the consequences of these policies, which alienated both the curas and the former mission Indians. However, neither curas nor royal governors advocated private property or an end to the communal production of temporalidades.[31]

In this period, Indian populations stabilized due to their acquisition of resistance to Old World disease and to a more equal sex ratio. The number of former mission Indians reached twenty thousand by the 1790s, and did not drop below that level again until the 1870s. Under secular Spanish rule, both curas and royal governors demanded more labor from the former mission Indians, but gave less in return than the Jesuits had. Some Indians reacted by migrating to Brazil, Santa Cruz, or Chuquisaca (modern Sucre), or else to savannas or forests free from Spanish control. For example, Orbigny noted in the 1830s that one thousand "still savage" Baures lived isolated near the Iténez River, and that some of them had left the missions after the expulsion of the Jesuits. Other Indians played the factions of curas and governors off against one another, and some cabildo leaders took decisive action. In 1793 and again between 1801 and 1822, the former mission Indians of Mojos resisted several royal governors by force under the leadership of the Canichana cacique Juan Marasa. Mojo leaders such as Pedro Ignacio Muiba also contested the authority of the governors. Migration to areas beyond Spanish control and political action under an experienced native elite were ways by which former mission Indians exercised agency, exploiting their dual heritage as participants in the Spanish colonial state and residents of the Amazon frontier.[32]

## The Former Mission Indians after Bolivian Independence

Although the activism of Marasa and Muiba became somewhat intertwined with the wars of independence (1807–25), those conflicts centered in the highlands and Santa Cruz and left the Llanos de Mojos virtually untouched. Upon achieving independence in 1825, Bolivia followed the French, Colombian, and Peruvian examples and replaced a centralist monarchy with a centralist republic divided into departments headed by appointed prefects. The initial departments took their names from the major colonial cities that served as their capitals, and as power centers of the local creole elites who replaced the Spanish. In the early years of the republic, the Province of Mojos fell under the administration of the Department of Santa Cruz, and Cruceño merchants and administrators exploited Indian production as the Cruceño curas had previously.

After independence, the Indians continued to produce goods, including cacao, cotton thread, cotton cloth, tamarind, coffee, vegetable oils, and wax, for the state under the temporalidades system. The Bolivian authorities continued to use cabildos to mobilize Indian commoner labor for the production of temporalidades as well as for road work, river transport, and mail-carrying services. As late as the 1840s, the Bolivian officials in the former mission towns received their salaries in temporalidades products. At least in the first half of the nineteenth century, the Bolivian national government only involved itself with appointing local officials and occasionally adjusting the cash value of temporalidades products. The Indians of Mojos supported their administrators with no help from the national government.[33]

Orbigny described one aspect of the temporalidades system as he saw it in 1832. Every fifteen days, each Indian woman was given twenty ounces of cotton, with seeds, which she had to have spun after another fifteen days. On the appointed date, the cacique placed himself at the door of the *colegio* (the old conventual buildings attached to the church) with a scale to verify that each Indian woman handed over four ounces of spun thread. The women sat in the corridor of the colegio to wind their thread, then returned to the cacique where they were called by name to verify the weight and fineness of their thread. Those that presented too coarse a thread were whipped. Every Indian woman received a piece of locally manufactured soap in exchange for her ball of thread.[34]

Orbigny implies that the temporalidades system had ceased to be the more or less equal exchange of goods that it had been under the Jesuits. Though the government continued to provide four hundred cakes of salt, two thousand pounds of iron, three hundred pounds of steel, four hundred knives, two hundred pieces of woolen cloth, sacramental flour and wine, and paper in exchange for temporalidades, the French visitor commented on "how insufficient such portions must be for a population of 23,000 souls." In the above description, spinners got only locally made soap in place of the salt, iron, and other exotic products that the Indians had received in exchange for their cacao and processed cotton. Still, some reciprocity survived as late as 1841, when the prefecture of Santa Cruz ordered the purchase of fifty quintals of iron, two quintals of steel, and six hundred cakes of salt, to be provided to the ex-missions of San Ramón and Huacaraje in exchange for their temporalidades remittances.[35] However, the national government's practice of remunerating its officials on the llanos with portions of temporalidades was an arrangement that invited corruption. In 1832, Orbigny noted that one reform-minded governor caught all of the town administrators keeping more produce for themselves than what they turned over to the state. Another cause of hardship for the Indians was that the demands of temporalidades production left scarcely fifteen days a year free for the planting and harvesting of food crops.[36]

## The Cultural Environment

Mojo society was quite complex, and people undoubtedly identified with several different groups. The largest sense of community was that of all former mission Indians of Beni, which Orbigny called the *naciones moxeñas*.[37] These people realized that they shared a similar culture, history, and religious values that set them apart from the irreligious, exploitative, immigrant Bolivian whites, as well as from the pagan, "uncivilized" indios bárbaros. At a more local level, former mission Indians belonged to a town, and relations between towns were not always amicable. The Canichanas of San Pedro and the Mojos of Trinidad (Trinitarios) maintained a longstanding enmity, to the point where San Pedro allied with the Mojos of San Ignacio (Ignacianos), against Trinidad and the Mojos of Loreto (Loretanos).[38] This enmity lasted at least until 1887, when Canichanas helped crush the Trinitario millenarian movement. At a still lower level, an individual Mojo might feel part of the familia or pueblo orders, of a parcialidad, or of an extended or nuclear family. The social divisions of familia and pueblo continued to exist at least as late as the 1830s, when Orbigny found that all of the Jesuit religious and administrative institutions were still intact.[39]

At this time, the number of whites living among the naciones moxeñas was still quite small. It should be remembered that "white," as elsewhere in Latin America, is a cultural more than a racial term. Bolivian whites (also called creoles in some sources) include individuals of mixed ancestry, but all live according to an essentially Spanish cultural pattern.

The Mojos call such people *carayana*, and although the term is already plural, non-Mojo speakers frequently use the plural "carayanas." The carayanas usually referred to themselves by the colonial term *vecinos* (town residents or householders).[40]

Though the Indians abandoned some towns and founded others in the late eighteenth century, most continued to live in the important Jesuit mission centers, some of which were relocated several times because of floods. The church and its plaza still formed the focus of each Indian town. The Mojo towns of Trinidad, Loreto, San Javier, and San Ignacio were dominated by their churches. Although none of the original churches survive, we know that they were similar to the earliest Jesuit churches of Paraguay, and to the Jesuit mission churches of Chiquitos in the Department of Santa Cruz, which still exist. They were a unique adaptation of European architectural ideas to the climate of the tierra caliente. The llanos churches were designed in the classic basilican plan of a nave separated from two side aisles by lines of columns supporting the roof. The columns were carved from massive tree trunks floated to the savanna from sources in the forests at the edge of the Andes. Each church had thick adobe walls protected from the rains by a huge, low-hanging, shallow-pitched roof covered with tiles. The roof extended out beyond the church facade to form a large atrium also shielded from the rains.[41] The Trinidad church's facade was decorated with frescoes of San Francisco and San Luis de Gonzaga, and its tympanum had a representation of the Holy Trinity as a single figure with three faces.[42] The church had three doors in the facade and one on each side wall, with the pulpit being placed between the side doors to take advantage of the breeze.[43]

Inside the llanos churches a good deal of the wealth of the Jesuit period remained, including organs, wooden statues, ivory crucifixes, and gold chalices, along with silver ornaments, vessels, candlesticks, lecterns, bells, reliquaries, and coats of arms. However, as the nineteenth century wore on, insects destroyed wooden images in some churches, and bats fouled reliquaries with their droppings. Innumerable bats roosted in the rafters of each town's church, and kept the air clear of mosquitoes when they flew out at night. One traveler who saw the bats leave the church of Reyes at dusk wrote that "there was a broad stream of them flowing across the plaza away toward the prairie. The flight continued for many minutes, and I am sure that the column must have been more than a mile in length."[44]

In Trinidad, the church and conventual buildings occupied one side of the plaza, and surrounded an interior patio which was separated from the plaza by a wall.[45] Soon after Trinidad became capital of the Province

*Figure 1.2.* Plaza of Exaltación in the 1860s. From Franz Keller-Leuzinger, *The Amazon and Madeira Rivers* (Philadelphia: J. B. Lippincott, 1875).

of Mojos in 1824, the two-story conventual buildings were converted into the government house, which later became the *prefectura* (prefecture or prefect's residence) with the creation of the Department of Beni in 1842.[46] To the other side of the church was an open, wood-frame bell tower, which reminded at least one visitor of scaffolding.[47]

In front of the church and its attached buildings was the great central plaza, the focus of the settlement. Usually beautified into public parks today, in the mid-nineteenth century the plazas were still just dirt and weeds, used as the ceremonial space for religious processions or the indigenous *machetero* dance, or as a parking area for the wooden oxcarts with two solid wheels which were common in the llanos. A large wooden cross, often elaborately carved, invariably marked the exact center of the plaza.[48] At least two former mission towns, Concepción and San Pedro, also had chapels placed at each of their plazas' four corners, used as stops in religious processions. Similarly, the plaza of Trinidad was marked by one cross at the center and one at each corner.[49]

The church and plaza were the center of Mojo ritual life, and in this respect represent continuity with the functions of the pre-mission bebedero and plaza. Orbigny noted that the naciones moxeñas devoted more time to Catholic religious practices than did the Indians of Chiquitos, and that the Mojos were the most devout of all the llanos natives. He observed that young people went to church in the morning and the afternoon for religious instruction. At 8:00 p.m. the entire community gathered to recite the rosary, and every Saturday they carried a religious statue in procession around the plaza, preceded by dancers (apparently macheteros).[50] Lardner Gibbon, an American naval lieutenant who visited the llanos in 1852, observed that the Mojos attended church every morning, and that the Mojo boys knelt and prayed before the central plaza cross every evening. In contrast, the local whites stayed up late gambling and slept late rather than attending church in the morning.[51]

Some dances represent continuity with pre-mission dances, with the most important being the machetero dance (also called *chiriperono* and *tontochis*). Machetero dancers (all male) perform in white tunics, wearing rattles around their ankles and an elaborate feather headdress with a tail of jaguar skin hanging from the back. They carry a wooden implement descended from a war club, but resembling a machete, which gave the dance its name. The machetero dance thus originated as a war dance, which the Jesuits converted into a "guard of honor of the Christian temple and doctrine."[52] The Jesuits transformed the pre-mission arrow dance, or

*Figure 1.3.* Ceremony on the plaza of a llanos town in the 1830s. From Alcides d'Orbigny, *Viaje a la América meridional . . . realizado de 1826 a 1833* (Buenos Aires: Editorial Futuro, 1945), reproduction courtesy Archivo y Biblioteca Nacionales de Bolivia.

*takiríkire*, into the *torito* dance, so called because its central figure wears a bull mask. Other figures wear other animal masks or represent satirical imitations of whites.[53] The machetero and the torito are both civic-religious dances performed for patron saints' days and Holy Week. Christmas-season dances (*danzas navideñas*) show more European influence than either the machetero or the torito.[54]

The plaza became a center of ritual and festivity during the town's patron saint's day. In Trinidad, this was Trinity Sunday, a moveable feast day occurring between May 17 and June 20, on the eighth Sunday after Easter. For the feast day of 1852, the Mojos set up arbors of green foliage and flowers at the four corners of the plaza, which they visited in procession after mass in the church. The prefect and clergy marched at the head of the procession, followed by the Mojos in their white cotton tunics (called *tipoys* for women and *camijetas* for men); some Mojos carried wooden religious images on their shoulders. Later, the Mojos took down the arbors and built an enclosure in the corner of the plaza in front of the prefecture, and held a bullfight there with "a number of savage bulls, wild from the pampa." Drunkenness was as much a part of the Christian feast day as it had been part of the ceremonies of the bebedero, though its meaning may have been different. The creole departmental officials put on a show of reciprocity by providing the Mojos with "great jars of chicha" and emptying baskets of corn and manioc bread into the crowd from the second-story balcony of the prefecture.[55] Of course, the Mojos did not need these "gifts" and manufactured both the chicha and bread themselves.

Holy Week was also an important time for ritual. In Trinidad in 1832, an aged Mojo volunteered to play the role of Jesus on Holy Thursday. Other Mojos formed a troop of "Jews" who escorted the Jesus impersonator, tied to a column, to the four corners of the plaza, where they whipped him. Good Friday was marked by church services, processions with religious statues, and flagellation by penitents. After a day of quiet on Saturday, the Indians celebrated on Easter Sunday as they ended their fasts. Families brought out chicha that they had prepared beforehand, and people became intoxicated more quickly after fasting. The celebrants also received an extra ration of meat. In Orbigny's time, Easter Sunday also marked the date when all of the town administrators converged on Trinidad to remit their temporalidades to the state.[56]

Originally, three sides of the plaza were surrounded by rows of long, rectangular dwellings, generally with palm-thatched roofs, called *cuarteles* (quarters or barracks). Their only furnishings were hammocks and large earthenware jars for storing chicha. Orbigny says that many families lived

MOJOS INDIAN, FROM THE ANCIENT MISSION OF TRINIDAD,
DANCING BEFORE THE ALTAR ON A HOLY DAY.

*Figure 1.4.* Mojo Machetero dancer in church in the 1860s. From Franz Keller-Leuzinger, *The Amazon and Madeira Rivers* (Philadelphia: J. B. Lippincott, 1875).

in a single cuartel, in rooms separated by simple partitions.[57] He does not say whether the families that occupied a single building were related, but it is likely that they were, and moreover that cuartel residence was ordered by parcialidad. The cuarteles were regularly spaced but did not form blocks equal in area to the plaza as was common in most grid-planned Spanish American towns.

Other important buildings included those of the *talleres*, or workshops, of the familia artisans. Trinidad had six workshops, three of which had their own buildings: one for weaving with twenty-seven looms, one for the thirty carpenters, and one for the eight blacksmiths. The ten shoemakers worked in the corridor of another building, the twenty brick and tile makers had two kilns and worked outdoors, and the six tanners also worked outside. With minor variations, all of the other towns of the Province of Mojos had similar facilities for their artisans.[58]

## Agriculture and Cattle Raising

The agricultural and cattle-raising arrangements of the Jesuit period continued with little change in the mid-nineteenth century. Beyond each town lay its agricultural lands, devoted to manioc, corn, plantains, rice, sweet potatoes, beans, sugarcane, cotton, coffee, and peanuts, among others. Cacao groves were located further away, since they required little attention. The type and location of cultivation varied from town to town according to the local environment. Originally communally owned, most of these agricultural properties would be broken up in the mid-nineteenth century in accord with liberal policies and gradually passed into carayana hands in the late nineteenth and early twentieth centuries.[59]

For most crops (except cotton and cacao), the Mojos continued to employ a shifting, slash-and-burn strategy of cultivation that was appropriate to their environment, with slash-and-burn plots being called *chacos*. Each crop had its own schedule for when to clear, burn, plant, and harvest. Clearing and burning had to be done in the dry season, but close enough to the beginning of the rainy season so that the crops would receive rain, generally in June, July, or August. Orbigny wrote that apart from biweekly rations of beef and a little fish, the standard Indian diet was composed of corn, manioc, plantains, squash, and forest fruits. Of the nonstaple food crops, Gibbon noted that cacao was gathered in November, coffee in May, and sugar in August and September. Despite the low relief of the region, different crops grew better at even marginally different altitudes. Thus rice was planted at lower elevations, sugarcane at middle elevations, and manioc and plantains at higher elevations, with corn being adaptable to different elevations at different times of the year. Both manioc and corn were made into flour.[60]

Beyond the circle of agricultural lands lay the savanna, which formed a vast unfenced pasture for the cattle of the Mojos. Each town's people referred to its pastures as its *campos* and marked their boundaries with those of neighboring towns. Most ex-missions had a number of outlying settlements called *estancias* where Indian cowboys lived with their families.[61] Mojo cattle raising in the mid-nineteenth century cannot technically be called ranching, since that term implies individual ownership of both animals and land, and the Mojos owned both animals and pasture communally. Like other people defined as "pastoralists," they killed fewer of their animals than ranchers do and produced mostly for subsistence where ranchers produce for the market.[62] The Mojo method of handling cattle was derived from Spanish practices transferred to the lowland regions of Spanish America. This lowland system focused on cattle and horses, with sheep unimportant; it was labor extensive, utilizing mounted cowboys who moved with the herds; it featured minimal supervision of semiferal cattle, with few roundups per year; and when it produced for the market, it produced hides and tallow in preference to beef. Though the Mojos let adult cattle roam freely, they penned the calves to protect them from jaguars and to obtain milk when the cows returned to nurse their young. The yearly floods created one local modification of Spanish lowland cattle culture: as previously alluded to, the Mojos would drive their cattle to the higher islas to escape the rising waters. When the waters receded, they would burn the grass to promote new growth, and the cattle would eat the shoots of young grass.[63]

We have already seen how the cabildos supervised cattle raising through an official called the alcalde de estancia. The cabildos also controlled the distribution of meat. On every second Saturday, the Indians killed fifteen to twenty animals, and laid out portions for each family on skins stretched out on the ground. The cacique and alfereces oversaw the distribution, giving meat first to married women, and then to widows, single women, and children. Each person went to get their meat, passing between two lines of fiscales armed with whips to maintain order. All of this took place in the open air, so that the recipients had to fight off vultures to retain their portion of meat. The normal ration of meat was doubled on a town's feast day, as well as on Easter Sunday. The Cruceño town administrators received fresh meat every two days, an indication of privilege. Important persons also brought meat along during their river voyages. Orbigny estimated that all of this consumption required the slaughter of five hundred to nine hundred head of cattle annually at each town.[64]

The provision of free beef to administrators and Indian families was part of a general system of food redistribution called the *mesa común*. Access to free food was especially important to familia artisans, who did not grow their own food. By the early nineteenth century, carayanas other than the administrators were helping themselves to the mesa común, so that Beni historian Manuel Limpias Saucedo could note that "the merchant to Mojos knew that in every town, the proverbial plate of rice with plantains and manioc awaited him, served and hot."[65]

## Conclusion

The Mojos and their neighbors were well adapted to life on the llanos at Spanish contact. After suffering the incursions of Spanish settlers from Santa Cruz, they evaluated their situation and eventually decided (in an early example of Mojo agency) that a Jesuit presence would both protect them from slave raids and provide access to salt and iron. In partnership with the Jesuits, the Mojos created a new, viable mission culture based on agriculture, artisan industry, and cattle raising, managed by the Jesuits and the indigenous cabildos. After the expulsion of the Jesuits, Mojos continued to exercise agency through rebellion and migration. And although new officials replaced the Jesuits, and the Bolivian republic succeeded the Spanish empire, the Mojos' mission culture survived remarkably intact. Little change occurred in native government, social structure, religious practices, and settlement patterns. Artisan production, agriculture, cattle raising, and river transport all continued much as in the mission era, except that the profits were co-opted by secular Spanish and Bolivian authorities for their own benefit. However, new ideas of government arrived with Bolivian independence, and before long, functionaries in the distant highlands began laying plans for the llanos that would bring the first major changes to the Mojos and their neighbors.

CHAPTER TWO

# *Liberalism Comes to the Llanos*

## Bolivian Rule and Liberalism

Although Bolivian independence brought little immediate change to the Province of Mojos, it stirred debate among national leaders regarding the constitution and government of indigenous communities. Liberal thinkers regarded the existence of corporate property and the separate legal status of Indians as barriers to economic progress and to the incorporation of Indians into national society. Ideally, this meant the abolition of both Indian tribute and of communal title to land, but for most of the nineteenth century, neither action was feasible. The Bolivian government initially had no source of revenue to equal tribute and recognized that Indians needed their communal land to be able to produce tribute.[1]

In sparsely populated Mojos, liberals would be able to enact reforms without threatening state revenue, which depended on the much larger indigenous population of the highlands. Mojos became a target of reform because President José Ballivián was interested in redefining the status of Indian communities and in furthering development and state control in the eastern lowlands.[2] He and his advisors were surely sensitive to the opinions of a cultured European observer like Orbigny, who had thoroughly investigated Mojos and judged it to be a place of "slavery and despotism under a free government," with the result that the "virgin land" was not even half as productive as it could have been. Orbigny pointed out that Mojos suffered because Santa Cruz held a near-monopoly on its trade, and was "considered, in a certain way, a possession of the inhabitants of Santa Cruz, strongly interested in hiding the slavery of the indigenes." The French visitor suggested that Mojos would progress if trade were opened with Chiquitos, Cochabamba, Brazil, and Europe. He believed that steamboats could pass the cachuelas and that Mojos could develop a modern iron industry.[3] These suggestions proved to be overly optimistic, but Ballivián clearly believed that breaking Cruceño political and economic control and alleviating the "slavery" of the Indians were necessary to unleash the productive potential of the llanos. Reform began in 1842 when Ballivián's liberal government separated Mojos from Santa Cruz on August 6, and then created the new Department of Beni on November 18. Unlike previous Bolivian departments, but like most French Revolutionary ones, Beni took its name from a river, a sign that Ballivián and his advisors were serious about imitating French liberal theories of government.

Liberal language filled the decree separating Mojos from Santa Cruz, which declared the post-mission regime to be "contrary to nature, to the Enlightenment, [and] to the constitutional principles proclaimed by the republic." The decree elevated the ex-mission Indians (who formed the vast majority of the population) to "the class of Bolivian citizens and as such, capable of the rights of equality, liberty, and property which the laws guarantee to Bolivians." It went on to state that individuals could now own land privately and that the Indians would now own the houses they occupied, but that the local government should construct separate houses for each nuclear family.[4]

Reform in Mojos was part of a nationwide trend, and the 1840s and 1850s were a high point of inclusionary, idealistic liberal theory in nineteenth-century Bolivia. Rival caudillos José Ballivián (1841–47) and Manuel Isidoro Belzu (1848–55) held the presidency, followed by Belzu's son-in-law and protegé Jorge Córdova (1855–57), and each leader tried to create a broader base of support. Ballivián's presidency produced the constitution of 1843, which specified employment (other than domestic service) as the only qualification for male suffrage, whereas previous constitutions had included an additional literacy requirement. Belzu introduced direct election of senators and representatives. Although much of his rhetoric on economics was antiliberal and protectionist, he promoted political action by Indians and the lower classes. Belzu's constitution of 1851 granted the exercise of political and civil rights to all males over twenty-one, and in 1855, he promulgated a resolution that specifically granted Indians suffrage and civil rights (although in practice, suffrage seems to have depended on literacy and an annual income of four hundred pesos from sources other than domestic service). Belzu also opened government employment to all, regardless of their capacity to vote. The constitution of 1851 remained in force until 1861, when a new constitution required literacy and either real estate or an annual income of two hundred pesos from sources other than domestic service for suffrage.[5]

In one way or another, most Bolivians would be excluded from voting until the mid-twentieth century, and the principal intent of Ballivián's and Belzu's reforms was merely to ensure political participation by artisans in highland cities. Nevertheless, the reforms marked an expansion of the bounds of citizenship within the limits of an oligarchic liberal state, and the 1840s and 1850s were the first time that indigenous people were exposed in a sustained way to a government language (if not a practice) promising equality and citizenship.

## Local Politics

The creation of the Department of Beni brought new officials to the llanos, including a departmental administration headed by a prefect appointed by the national government. After a failed plan to establish the capital at a new city on the Beni River to be called Ciudad Ballivián, and a subsequent attempt to make Exaltación the seat of government, the prefect took up residence in Trinidad. The Bolivian government introduced the office of corregidor as its local representative in each former mission town, and each town's district became known as a *cantón*. The prefect appointed the corregidores, who were most often whites, though there were Indian corregidores in the Mojo towns of Trinidad and Loreto and in the Cayubaba town of Exaltación. As the highest-ranking Indian official in Trinidad, the corregidor was often considered to be head of the cabildo, but technically he was not part of it. His authority was circumscribed by a carayana municipal council created in 1849, which restricted the corregidor's duties to governance of the Indians.[6] Another, relatively low-ranking office open to Mojos was that of *alguacil del juzgado de letras* (bailiff of the court) of Trinidad. A Mojo, Manuel de la Cruz Mobo, served as bailiff in 1853–56, and another Mojo, Patricio Cayuba, succeeded him.[7]

In their management of temporalidades, the corregidores could be as corrupt as the administrators of Orbigny's day. Bolivian historians Zulema Lehm, Arnaldo Lijerón, and Lorenzo Vare mention that corregidores were supposed to remit all produce from communal labor under the temporalidades system to the state, but that "a great part of that production was 'illegally' commercialized by the corregidores for their own gain." These officials consequently pressured the cabildos and the Indian populace to produce more goods, so that more could be skimmed off.[8]

The former mission Indians continued to be politically active, and Indian leaders found room for independent action by exploiting the chaotic nature of carayana politics. When highland military leaders fought over the Bolivian presidency, they sent their allies to Beni as prefects, and their enemies there as exiles. As a result of this exported factionalism, several Beni prefects lost their positions in coups and reclaimed them in countercoups between 1848 and 1873. Many of those who fought over the prefecture relied on Indian support, and leaders like Frutos Nosa (Mojo corregidor of Trinidad), Domingo Avaroma (Cayubaba corregidor of Exaltación), and Pedro Pablo Tonore (Movima cacique of Santa Ana) took full part in the power struggles, undoubtedly with their own interests at heart.[9]

Frutos Nosa was born in 1796, and was thus in his late forties during the first years of reform. Married with several children, he served as corregidor in the years 1843–45, 1848, 1852, and 1854, and as cacique (cabildo head) in 1847, 1862, and probably other years. Gibbon, who called him "Fratos," described him as "an old Indian, [who] is considered the rich man of Trinidad," and wrote that he held his office because of his wealth. When Gabriel José Moreno deposed Prefect Rafael de la Borda in January 1848 in the turmoil following the overthrow of Ballivián, Nosa supported the coup and addressed the Mojo inhabitants, securing support for the new government. Nosa was still active in local politics in 1864 and 1865, when he joined the carayana vecinos in taking action first to oppose, and then to accept, the government of President Mariano Melgarejo.[10]

Most indigenous participation in carayana politics tended toward an opportunistic siding with whatever faction seemed most powerful or likely to win. A prime example is when the exiled lieutenant colonel Atanasio Baldivia rose up against Prefect Dr. Francisco Ibáñez in November 1848 in support of Belzu's coup d'état. The carayana corregidor of Loreto prepared to resist Baldivia with sixty carayana riflemen and fifty Canichana and thirty Guarayo bowmen, but then surrendered, his Indian and white followers subsequently joining Baldivia's forces. Baldivia allowed Ibáñez to go to Santa Cruz, but the latter returned to Trinidad in December with three army officers and twenty-five national guardsmen, intent on retaking power. Loretanos, Trinitarios, and several vecinos of Trinidad joined Ibáñez, and during a battle against Baldivia, the Canichanas and Guarayos passed to Ibáñez's side too. Having taken Baldivia prisoner, Ibáñez's largely Indian forces proceeded to loot the houses of Trinidad's vecinos. Indian participation was crucial again in January 1854, when Ramón León Carranza overthrew Prefect Dr. Manuel Lea Plaza. Carranza threatened to shoot Lea Plaza if the Indian masses (*indiada*) rose up in his support, and then sent Lea Plaza toward Santa Cruz by boat. Two Mojos inspired the boatmen to switch sides, one being Patricio Cayuba, who was soon to become bailiff. They helped Lea Plaza overpower the soldiers guarding him, and the vecinos and Mojos of Loreto then supported his return to Trinidad. Carranza fled to the Iténez and died in an accident before a large force of Canichana bowmen sent by Lea Plaza could arrest him.[11]

In certain extreme cases, former mission Indians would even kill carayana officials. Pedro Pablo Tonore and a Movima crew drowned outgoing prefect Luis Valverde while taking him to exile in Brazil in 1851. Limpias implies that the Movimas acted at the instigation of General Manuel Carrasco, who had just deposed Valverde, but Gibbon states that Valverde met his death because he "was exceedingly tyrannical in his behavior to these people." Similarly, when the Ignacianos killed their carayana corregidor in 1855, it may have been due to the excessive demands he made on boatmen.[12]

Alongside the new officials of the civil government, more traditional representatives of spiritual government continued to serve. The Catholic Church remained the official religion of Bolivia, and each former mission town had a priest who received a government stipend. The ex-mission Indians had great respect for the church but would not support an abusive priest without taking some kind of action. When Manuel Angel del Prado, the bishop of Santa Cruz, made a pastoral visit to Beni in 1850, the Itonama authorities of Magdalena denounced their priest. They claimed that Father Pedro Pablo Salas had left the dead unburied for three or four days, that he had mistreated Indian sacristans, and that he had let people die without receiving the sacraments and then stolen their goods. Prado's secretary recorded the testimonies of several prominent Itonamas, but it is unclear if Salas was ever punished.[13]

## Reform of Property and Taxation

Liberals in republican Bolivia saw Indian communal lands and other vestiges of the colonial era as barriers to progress and tried to do away with them. As mentioned, a liberal agenda informed the supreme decree of August 6, 1842, which separated the Province of Mojos from the Department of Santa Cruz prior to the creation of the Department of Beni. In the decree, the national government recognized the former mission Indians as citizens and promoted the idea of private property by ordering the division of land and the construction of separate houses for each nuclear family within the year, and conceding full ownership to the individuals who received such property. Those who received property would henceforth pay a head tax (*contribución de capitación*) of two pesos per year, while those who didn't receive property would pay one peso. Governor Miguel de los Santos Rivero distributed some land to the Indians and also gave individual title to several urban houses. New prefect José de Borja advised a return to the remnant of the Jesuit communal system upon his arrival in Beni, and the national government concurred in a supreme order of January 12, 1844. Then, in a new supreme order dated April 20, 1847, the Ballivián government again decided that communal property should be gradually replaced by private property, and commanded that lands and livestock begin to be divided by giving allotments to ten to twelve Indians from each town per year, starting with the hardest working and "most civilized." Other provisions required that artisanal equipment and state-owned boats be divided among private owners, that employers should always pay a wage to Indian laborers, that all indigenous men between the ages of eighteen and fifty pay a tax (called *impuesto personal*), and that those who received allotments of property pay both that tax and another (called *contribución*) on the property. In 1848, Prefect Ibáñez divided 2,496 cattle among fifty-two family heads, 4,320 cacao trees among fifty-four families, and 200 cotton plants among the poorest families.[14]

Liberal reform in Beni thus sought to replace communal property with private property and to establish new taxes. Nevertheless, the influence of colonial-style practice is evident in that the impuesto personal was essentially a tribute administered in the same way as among highland Indians; the former mission Indians would henceforth be counted in censuses (*padrones*) "by the method established for indigenous tributaries" so that the state could keep track of this income, and all people would be classified as either *originarios* native to their district, or immigrant *forasteros*. A document from Reyes indicates that the authorities were already dividing some ex-mission Indians into colonial-style tribute categories by 1845. The fully developed system labeled originario males as *niños* (to age fourteen), *próximos* (fourteen to eighteen), *tributarios originarios* (eighteen to fifty), *reservados* (over fifty), and *ausentes* (absent). Females were grouped as *niñas* (to age twelve), *casadas* (married women), *solteras* (single women), and *viudas* (widows). Of all these categories, only tributarios originarios had to pay the tribute of two pesos. The one detail that differentiates the impuesto personal from a pure colonial Indian tribute is the fact that forasteros and government employees also had to pay it, at least initially. Only its equal application to men of all races qualifies the impuesto personal as a "liberal" reform.[15]

During Belzu's administration, antiliberal in its economic policy, the national government took no further action against communal property. Then, by a supreme order dated June 7, 1856, the Córdova government ordered the final division of the communal cattle and horse herds among individual Indian owners, who henceforth had to pay an annual tax on each animal. On November 3 of the same year, the government dictated a supreme resolution abolishing most Indian communal lands in Beni, mandating that cacao groves, sugarcane fields, and tobacco fields be divided. Orchards belonging to the former mission towns were to be auctioned off, but cotton fields and sugar mills were to remain communal property for the time being.[16] The state took no interest in dividing corn or manioc plots at this time, for though corn and manioc were staple food crops, they had little monetary value compared to cacao, sugar, or tobacco. In addition, they were grown on slash-and-burn plots that were abandoned every few years, a system which would have been hard to accommodate in any formal division.

## Reform of Production and Labor

The Bolivian government reported a significant decline in the monetary value of products from the Province of Mojos between 1835 and 1838, when the total value of textile, agricultural, and artisan products dropped from 35,079 pesos to 16,011 pesos. Most of the decline reflects a decrease in the production of the cotton textiles called *macanas* from 26,740 ¼ varas worth 23,398 pesos in 1835 to 8,668 ½ varas worth 7,585 pesos in 1838. The government report noted that President Andrés de Santa Cruz's 1837 decree of free trade, which reduced tariffs on foreign textiles, only took effect in 1838 and could not explain this "crash" (*quiebra*). The general administrator of Mojos, Colonel Anselmo Villegas, alleged that bad harvests were the cause of the decline, though the government report stated that internal customs records of goods moving out of Mojos contradicted the reports of a real decline in production.[17]

Whether the decline was real or invented by Villegas to cover up illegal commerce, the Bolivian authorities were alarmed and soon tried to restore production in the ex-missions to the high levels achieved from the colonial period to the early 1830s. In 1841, the national government ordered the prefect of Santa Cruz to send a young native of Mojos to Cochabamba in order to learn how "to establish and direct cotton-weaving factories." The prefect repeated the order to the governor of Mojos several times without result, and eventually gave up the attempt while scolding the governor.[18]

As the persistence of the temporalidades system ran counter to liberal ideas of a free market, the Bolivian government began to enact legislation against it after separating Mojos from Santa Cruz. The decree of August 6, 1842, specifically declared Indian women "free of all forced labor or personal service in favor of the state." The authorities thus relieved the Indians of the obligation to produce cotton cloth for the state, believing that production would increase in a freer economy. In 1843, the national government ordered the prefect of Beni to revive the production of fine textiles, declaring that the industry "will be stimulated by the advantages that free trade will offer to manufacturers." Accompanying the order were three sample books of old cloth made in Mojos, which were to serve as examples. Prefect Borja responded that textile production had declined and that the types of cloth represented in the samples were no longer made. Since cloth production was no longer obligatory, Borja planned to persuade the Indians of the value of imitating their predecessors as a way to protect their liberty and property, and would explain this to the caciques, cabildos, and "weavers' guilds" (parcialidades) during his official *visita*. He would also encourage more planting of cotton to replace fields abandoned by the administrators when state cloth production was discontinued. By 1844, Borja was able to send the minister of the interior examples of new cloth made by the Indians in imitation of the colonial samples, and to claim that the entire industry would be revitalized within a year. The prefect also stated that the natives had made the new cloth as a payment of tax (contribución), indicating that their obligation to produce textiles for the state really continued under a new name. In the next year, Borja noted that the Indians were also producing textiles for their priests, whom the government had named administrators of the cotton fields in 1842. The state continued to receive free cotton as well, accepting one and a half arrobas of cotton as equivalent to the two-peso impuesto personal.[19]

The state again attempted to regulate textile production under Belzu's first prefect, Colonel José María Aguilar. His order ending the forced labor of female indigenous weavers in 1849 only proves that such labor tended to persist despite all government pronouncements. Prefect Valverde (1849–51) attempted to improve production by naming the artisan Miguel Forero director of Trinidad's weaving shops. Forero brought in eight new looms that allowed the Mojos to weave in a more comfortable position, with the result that macana production rose from two to forty varas per day.

Prefect Ricardo José Bustamante (1852–53) tried to simplify and maintain a high level of production by ordering that only three types of cloth be made. Gibbon noted in 1852 that Mojo women were not paid for picking cotton, spinning, and weaving, euphemistically calling them "volunteer workers." He also reported that Beni had exported somewhat more than three thousand varas of macanas and seven hundred cotton tablecloths in 1851. By 1856, it was clear to reforming Prefect Dr. Carmelo María del Rivero that the ex-mission Indians were submitting the entire year's worth of unpaid production as payment of the impuesto personal, and that both cabildo officials and carayanas were using commoners to do agricultural work without pay.[20]

Other agricultural production remained under the temporalidades system, and the departmental government went so far as to draw up regulations for agricultural tasks by month in 1844, adding that each town's administrator and cacique would jointly determine the amount of time the Indians could spend working their own chacos.[21] By 1845, the wording of two Bolivian laws regulating temporalidades suggests that the national government by that time regarded it as merely tax (contribución) paid in agricultural products rather than in money, with no mention of any obligation of the state to provide anything in return. In a further attempt to create a liberal-inspired market economy in Beni, the national government removed cacao from the list of temporalidades products in 1845, ordering that it be reserved for the Indians' "commerce with those who extract it from the province."[22]

To try to end the abuse of production by local government officials, the national government declared that the ex-mission Indians were to pay their two-peso tribute in coin rather than produce in the supreme order of June 7, 1856. Finally, on July 9, 1856, the government promulgated a decree that ordered an end to the entire system of temporalidades and created taxes on cacao and cattle in its place. This attempt to substitute cash payments for temporalidades products was an effort to replace colonial economic systems with liberal ones, in which people were to be free to sell their produce at whatever price the market would bear. However, the decree required the Indians of each cantón to perform unpaid, rotational service for prefects, governors, and priests in compliance with President Santa Cruz's decree of July 2, 1829, which had mandated this service for highland Indians.[23]

Because of the recent abuse of the temporalidades system, the former mission Indians greeted its abolition with joy. In Trinidad, after the cacique and other interpreters from the cabildo explained the decree to their people, the Mojos went to kiss the prefect's hands, and then retired to celebrate with chicha and cane liquor. Yet Lehm, Lijerón, and Vare write that although the government legislated its abolition many times, the temporalidades system came to an end only with the depopulation of towns during the rubber boom. Indeed, the government reestablished communal labor in December 1857. Prefect José Matías Carrasco, who had previously supported all liberal reforms, agreed in 1858, saying that the Indians had proved incapable of individual initiative and that "the results have been: collapsed buildings, lack of food supplies, and dispersal of cattle." Carrasco claimed that communal work in construction, repair, agriculture, and cattle raising would benefit the Indians, whereas the "refined feudalism" of temporalidades had benefitted others.[24] After the official suppression of temporalidades in 1856, local officials could presumably dispose of the products of continuing communal labor as they saw fit. Under the new system, corregidores enjoyed economic benefits that caciques did not, as the former kept part of all taxes collected for the government. These included taxes on cacao exported from Beni and on cattle captured from wild herds, the annual tax paid by landowners, and the annual tribute (impuesto personal) now paid by former mission Indians.[25]

One form of labor, service in river transport, was outside the temporalidades system and became the subject of separate reforms. Indigenous boatmen were essential for river travel in Beni, and the authorities hesitated to restrict carayanas' free use of their services. In 1843, Prefect Borja created new regulations for the state expeditions that carried mail from Beni toward Santa Cruz and Cochabamba. Captains were to be respected cabildo members capable of keeping the crew "perfectly subordinated," and they were to be held responsible for all infractions. In the interest of speed, Borja prohibited crews from stopping to hunt or fish, and ordered that they be provided with jerky in the port of Loreto rather than waiting for fresh beef. He also stipulated that nine towns rotate river service by month.[26] The supreme order of April 20, 1847, called for the distribution of state-owned boats among individuals "capable of subordinating the paddlers," who had to continue serving state needs for free. Prefect Aguilar ordered an end to the practice of using indigenous boatmen without pay.[27] Gibbon was told in 1852 that Indian paddlers were customarily whipped in Trinidad if they took too long on their voyage, and that the threat of such a whipping would motivate his own crew to paddle faster. His paddlers felt cheated by a local official, who would have travelers pay him for Indian paddlers in silver coin, and then pay the paddlers with cotton cloth.[28] The fact that whites spoke so often of the need to subordinate Indian crewmen indicates that the paddlers actively defended their interests.

In 1858, Prefect Carrasco promulgated a lengthy set of regulations governing river travel. Merchants were to make contracts with and pay indigenous boatmen in coin or goods in the presence of the authorities. The regulations stipulated the number of men and amount of cargo allowed for different-sized boats. Depending on which town an expedition left from, the new rules specified several different pay scales (ranging from two reales to three pesos) and the different amounts of fresh beef or jerky that merchants were required to provide the crewmen. Payment going downstream to Beni was to be half of the payment going upstream toward Santa Cruz and Cochabamba. Captains were to make one-quarter more than crewmen on long voyages and double on short, town-to-town trips. Captains had to attend to, prepare food for, and generally obey merchants and protect their cargo from water. Other regulations limited the amount of work expected of crewmen in loading and unloading. Two longer portages, from Reyes to the Yacuma River and from Loreto to its port, would always require an additional payment.[29] Although these exact regulations may not have lasted long, one thing is clear: Indian boatmen were paid for their work.

Cattle represented one of the principal sources of wealth on the llanos. But in the mid-nineteenth century, the livestock economy of Beni was less healthy than it had been in Jesuit times, or even during Orbigny's visit in 1832. The national government considered wild herds of cattle to be its property and issued warrants to round up and slaughter large numbers.[30] The entrepreneurs who received these warrants did not always distinguish between truly wild herds and the semiferal herds of the Mojos. Cacique Mariano Cayuba complained to Gibbon: "We used to have plenty of cattle and fine horses. The white man comes from Santa Cruz and drives them all away."[31] Even prefects took part in plundering the herds. Before leaving office in 1852, Fabián Hoyos sent 377 cowboys out of Loreto to round up a herd of cattle that was never seen again. Horses, essential to cattle raising, were also requisitioned occasionally.[32]

The reform of the mission system of production had many consequences, not least of which was an end to the last remnants of mission-style redistribution. The provision of salt and iron in exchange for temporalidades was already uncertain by the 1830s, and this reciprocity ceased altogether after 1842. In August 1843, Prefect Borja complained to the minister of interior about the lack of salt and the high prices charged by merchants from Santa Cruz and Cochabamba, and suggested that the Province of Chiquitos could exchange salt for some of Beni's products, a proposal that appears to have been put into effect. But soon, salt could only be obtained by purchase, and Gibbon found salt cakes from the Altiplano to be the principal item traded to Beni by Cochabambino merchants in 1852. In another restriction of redistribution, the minister of the interior ordered the suppression of the mesa común in September 1843, as Ballivián considered it an "obstacle to industry" and "burdensome to the natives." Borja acted on this order early in 1844, ending the supply of free food to both carayanas and familia artisans. Finally, the breaking up of the Mojos' communal herds in 1856 brought an end to both the cabildos' distribution of meat and to the office variously known as alcalde, teniente, or mayordomo de estancia.[33]

## Reform of the Urban Plan

Among his other reforms, Ballivián decided to modify the actual physical appearance of the former mission towns. Yet this reform of the urban plan was motivated not so much by liberalism as by a desire to make the mission towns conform to the Spanish and indigenous towns of the highlands, which were laid out on a grid of streets. Since the sixteenth century, the rulers of Spanish America had considered a grid plan, single-family houses, tile roofs, and street-front porticos to be among the essential physical manifestations of civilization, and had imposed them on many sedentary Indian communities.[34] In the eyes of the national authorities, the Jesuits had only half-civilized the Indians of Beni. Although the mission churches and plazas conformed to Bolivian urban ideals, the thatch-roofed, multifamily cuarteles had to be destroyed to complete the civilizing work begun with sedentarization and Christianization. The ideological reasons for such radical modification of the layout of the former mission towns thus reflect a discourse of civilization and barbarism that antedated, but was incorporated into, the newer creole discourse of liberalism.

As part of its attempt to dismantle the communalism of former mission society, the national government ordered that the cuarteles be replaced by single-family dwellings in 1842, and Prefect Borja began the reconstruction of Trinidad by July 1844. The Mojos did the work, directed by their corregidor, Frutos Nosa. By September of that year, Nosa reported that workers had finished building twenty-nine two-story houses with tile roofs on three sides of one block, had put new tiles on the roofs of some houses fronting the plaza, and had built a new, tile-roofed building for the tanners. By December, they had also fixed parts of the church, made three thousand adobe bricks, and gathered enough material for building up three sides of another block, including thirty wooden pillars with their capitals. In San Javier, workers had made sixteen thousand tiles and forty-two hundred bricks by December, in preparation for new construction. In other outlying towns, workers also prepared many thousands of tiles and bricks, but new construction did not always conform to the "civilized" ideal. San Ignacio's workers completed many houses on the new grid of streets, but these were almost all *tabique* (wattle and daub), with thatched roofs, and some were cuarteles rather than single-family homes. Loreto built a new cuartel of tabique and tile, while San Ramón built two cuarteles of adobe and tile. In many other towns, workers built new cuarteles or refurbished old ones, usually using tabique and thatch.[35]

Much of the new construction, especially outside of Trinidad, reflected local power structures and benefited civil, spiritual, and indigenous authorities. San Javier's workers prepared material for construction of the priest's residence. In San Pedro, workers concentrated on rebuilding the church and colegio, and on building a new brick and tile house on the plaza for the cacique. In San Joaquín, workers built a new house for the cacique and a new church bell tower of adobe and tile, as well as a house for the carayana administrator. In Carmen, they built an adobe and thatch house for the cacique, a tabique and thatch house for the cabildo *intendente*, and a room for the church-run school. San Borja built a house for the corregidor and one cuartel, both of tabique and thatch. Baures built thirteen new cuarteles, of which only two had adobe walls and one a tile roof, suggesting that the better construction was reserved for high-ranking parcialidades. The Baures also replaced the thatch roofs of four houses with new ones of tile.[36]

In many towns, whitewashing helped give a "civilized" appearance to the most important buildings, and sometimes the entire town. In San Pedro, workers painted part of the government house, including painting the national coat of arms on the wall. The Ignacianos whitewashed the colegio, the buildings facing the plaza, and some of the new houses, and the Loretanos whitewashed and painted their church, public buildings, and houses. The people of Santa Ana painted the exterior and interior of all their houses, and those of San Joaquin whitewashed and painted the church and the rest of the town. The Trinitarios managed to put their own imprint on the new urban plan by decorating the whitewashed walls of the market, together with those of their houses and of the rooms of the prefecture, with paintings of cattle fleeing burning grass, and of white men, cows, and chickens.[37]

Urban reform continued after Borja left office. In 1845, Prefect Rafael de la Borda directed the corregidores to continue the construction of new houses. In 1851, the Trinitarios constructed an *hospicio*, or lodging for visiting merchants, and began to build a hospital and renovate the prefecture. In 1852–53, work continued on two blocks in Trinidad, and many old tabique and thatch houses had to be rebuilt after the flood of 1853. By the 1850s, Trinidad had blocks with one-story buildings fronting the streets; the blocks had empty spaces in the middle, one of which hosted the market. On the street side, the roofs of these buildings extended over sidewalks and were supported by lines of wooden pillars. This style of architecture, still common in Beni towns, enables people to walk through town with a minimum of exposure to the rain. The built-up area of these towns was quite small: in Trinidad as late as 1920, it stretched only about two blocks away from the plaza in all directions.[38] Although most outlying towns did not immediately create all the requisites of a "civilized" Spanish town (Baures and Magdalena were still

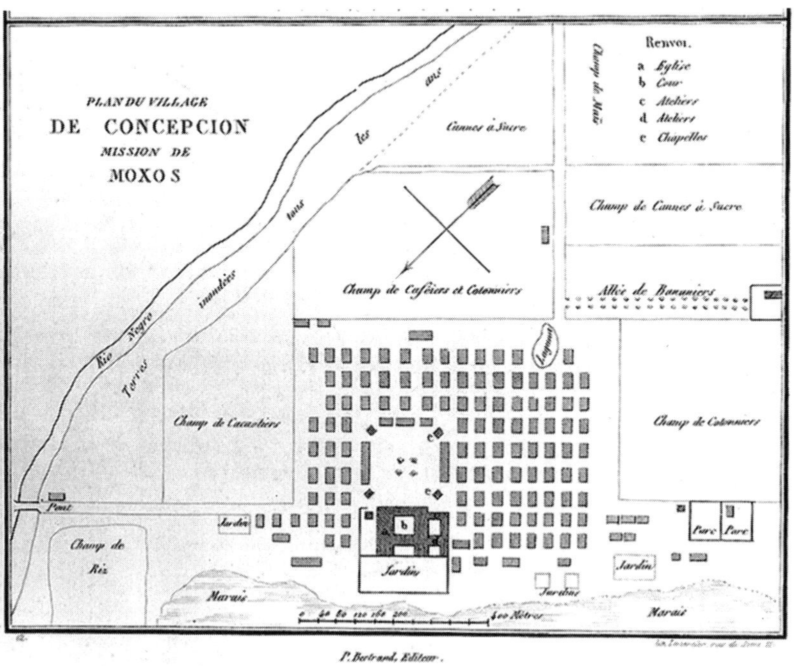


*Figure 2.1.* Plan of an ex-mission town before the liberal reforms: Concepción de Baures in the 1830s. From Alcides d'Orbigny, *Viaje a la América meridional . . . realizado de 1826 a 1833* (Buenos Aires: Editorial Futuro, 1945), reproduction courtesy Archivo y Biblioteca Nacionales de Bolivia.

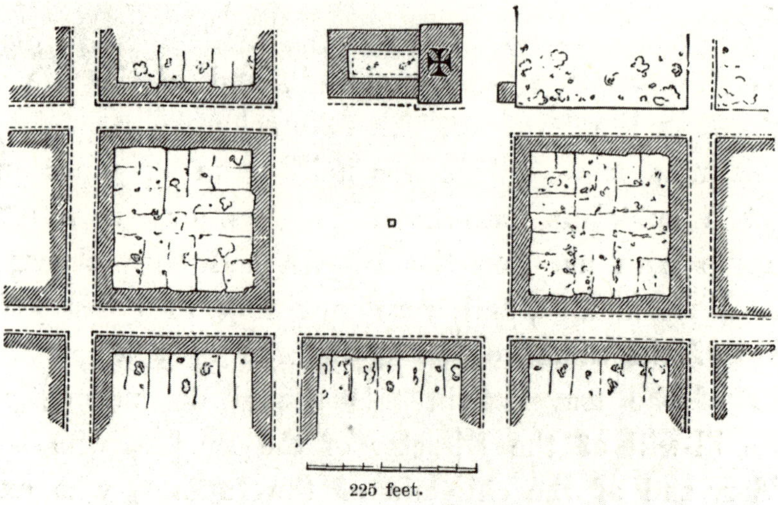


*Figure 2.2.* Plan of an ex-mission town after the liberal reforms: Exaltación in the 1860s. From Franz Keller-Leuzinger, *The Amazon and Madeira Rivers* (Philadelphia: J. B. Lippincott, 1875).

building cuarteles in 1852), the grid plan that had first replaced the rows of cuarteles in Trinidad slowly spread to other Beni towns. Exaltación built half a block of houses and a new school in 1853. That town had a grid plan of streets and houses with continuous verandas supported by wooden columns like those of Trinidad by 1868, and about a hundred adobe-walled, tile-roofed houses arranged on these blocks by 1874.[39]

Reform-era prefects also ordered other public works. In 1844, the former mission Indians made improvements to their cattle estancias "for the comfort of the cowboys and their families," constructing or renovating houses and cuarteles, fixing up old corrals and building new ones, and building or restoring chapels. San Ramón, Magdalena, Santa Ana, and San Ignacio created new estancias. The Ignacianos began theirs at a place called Jesús, but later had to abandon it "because of the many bats which then took possession of that place." Most estancia buildings were of tabique and thatch, but the Trinitarios replaced the straw roof of the estancia of San Lorenzo's chapel with one of tile. In the same year, each town worked on improving existing roads to their estancias or to nearby towns, clearing vegetation and building bridges and earthen causeways. Loreto graded its plaza, and Exaltación built a causeway 155 varas long to connect the town to the

Mamoré River. Later, in Borda's prefecture, officials measured the Laguna Socoreno near Trinidad and determined that it could provide water to the town in the dry months of August, September, and October. Mojos under Frutos Nosa's direction then built a causeway sixty-three hundred varas long between Trinidad and the lake, with ditches on each side to bring water. In 1851, the Trinitarios began another causeway to their river port of Trapiche. In 1855, natives of Baures, Magdalena, San Ramón, and San Joaquín dug a canal connecting the Mamoré and Ipurupuru rivers, providing a shortcut to the Iténez region.[40]

## Population, Citizenship, and Education

Population figures for the Llanos de Mojos in the mid-nineteenth century are widely divergent, ranging from 22,742 (Block's average for 1826–67) to 114,922 (the Bolivian census estimate for the Department of Beni in 1854, including areas now in the Department of La Paz). Several other figures from 1844 to 1852 are between 30,000 and 50,000, and we may accept this range in lieu of anything more definite. Almost all of the population consisted of the descendants of mission Indians.[41] Government tribute records from 1849 list 2,483 originarios in the four Mojo cantones, out of a total tributary population of 6,676 in the Province of Mojos, at that time roughly coterminous with the modern Department of Beni.[42] Orbigny cites total populations of 2,004 for Trinidad, 1,876 for San Ignacio, 2,145 for Loreto, and 1,515 for San Javier in 1832, almost all Mojo.[43] Bishop Prado found 5,000 natives and 150 forasteros in Trinidad, and 2,446 natives and 30 forasteros in San Ignacio in 1850. In other estimates of the population of Trinidad, geographer Wolfgang Schoop cites a figure of "scarcely more than 2,500" for 1842, while Gibbon says there were 3,000 people in 1852. Again, almost all of this population was Mojo.[44]

Bolivian liberals encouraged white immigration to the Llanos de Mojos by giving non-Indians access to former communal property in the decrees of 1842, 1847, and 1856. Partly as a result of this policy, the recorded white population increased from 57 to 1,191 between 1830 and 1855 (still a small percentage of the total population).[45] Most were from the lowland city of Santa Cruz, but a good number came down the Andes from Cochabamba. Many were merchants, but others were government officials. The prefecture was a reward given to allies of a sitting president, but appointment to a lesser post could serve as internal exile for political undesirables from the highlands.[46]

What did it mean to be a "citizen" under the new regime? The 1842 decree specified that Indians were citizens, and documents from 1843 and 1844 identify prominent cabildo members from several Beni towns as citizens, including the corregidor Frutos Nosa. The bailiff Manuel de la Cruz Mobo also called himself "citizen" in 1856, and the *revisita* of 1861 lists several Trinidad cabildo members as citizens. However, other Indians, and even a couple of white corregidores, are not called citizen in the 1843–44 documents. Local whites also used the liberal term *citizen* (*ciudadano*) and the colonial term *notable resident* (*vecino notable*) interchangeably. Frutos Nosa was considered both vecino and ciudadano as late as 1864–65, when he signed documents as an equal of the many carayanas who signed.[47] Whites thus interpreted citizen as meaning a local person of high status and translated liberal language into colonial-style practice.

Citizenship implied the right to hold public office, but in some instances Mojo officeholders found that they could not retain their positions. Frutos Nosa resigned as corregidor in 1845, "not considering himself of the necessary aptitude for the performance of the administration of the interests of the state that had been confided to his care." Presumably, Nosa found that his organization of labor for the intense construction campaign in and around Trinidad was either not worth his corregidor's salary or was making him unpopular among the Trinitarios. The newly arrived Prefect Borda wrote that Nosa was well suited to be cacique, but "absolutely lacks the indispensible capacity to command and administer with good success," and had been elevated to corregidor "without discernment, temporarily." Perhaps Borda made Nosa resign as part of a wider reevaluation of the officials appointed by Borja, a scenario that might explain Nosa's support for the overthrow of Borda three years later.[48] The bailiff Manuel de la Cruz Mobo also could not maintain his post, resigning in 1856 after three years of service. In his letter of resignation, he stated that he had barely been able to support his large family with his salary of fifty pesos per year and the produce of his own farm plot. A new regulation that his salary be paid by the treasury of distant La Paz meant he could no longer count on that money, and his duties as bailiff were causing him to neglect the crops on which his family depended. He therefore reluctantly resigned his position, and ended the letter with a reaffirmation of his sense of citizenship, "always offering, for other occasions, all of my patriotism and service as lover of the constitutional regime."[49]

The citizenship of Indians remained open to debate in Bolivia until 1938, but nineteenth-century politics limited the participation of both Indians and non-Indians. According to historian Rosanna Barragán, Bolivia's nineteenth-century constitutions recognized a difference between Bolivians and citizens. All people born in Bolivia were Bolivians and were entitled to civil rights, but only those who met literacy and employment requirements were citizens with the right to vote and hold public office. Fewer than 10 percent of adult males voted in Bolivian elections up to 1952, and the figure was usually less than 5 percent. Barragán has calculated that only 1.3 percent of Beni's adult males voted in the election of 1844, and 0.5 percent in that of 1854. She has also observed that, despite Bolivia's largely rural society, nineteenth-century electoral politics took place mostly in cities, and above all in departmental capitals.[50] Within Beni, liberal policies and ideas did indeed have the most impact in the departmental capital, Trinidad. Only there did carayana authorities continue to refer to some Indians as citizens into the 1860s. While Trinitario leaders were content to be recognized as equals in this period, there is as yet no evidence that they promoted the idea of citizenship among commoners. Nor do commoners seem to have reclaimed their rights as citizens. The language of liberalism was still too new, and both whites and Indians simply translated the new language into colonial-style practice.

Evidence of strained relations between carayanas and Mojos also belies the rhetoric of equality. The Mojo cacique of Trinidad, Mariano Cayuba, told Gibbon "we are all slaves to the white man," and when Gibbon visited Loreto with the prefect, he found that one Mojo left for the countryside because he disliked "all the creole race" and another refused to let his daughters marry creoles. Gibbon summed up Mojo feelings by writing that "the Indians of Mojos are not friendly to the Spanish race at heart," but added "that they love and respect the influences and arrangements of the church there is no doubt."[51]

Nevertheless, former mission Indians and carayanas all could claim to be civilized in the traditional, colonial sense: they were Catholics, lived in towns, and were not bárbaros. Bishop Prado told the Baure authorities of San Joaquín to act as moral examples for their people, to achieve "the advancement of their town, as much in political as in moral and religious civilization." He also advised the Itonamas of San Ramón and the Baures of Carmen to end the "bad habits" and "bad inclinations, which have their origin perhaps in [the time of] their paganism." Prado thus conceived of Christian civilization as something tenuous and uncertain that had to be promoted among the ex-mission Indians on this frontier.[52] The civil authorities also maintained a traditional attitude toward "bárbaros." Prefect Borda wrote in 1846 that they frustrated attempts to develop Beni, either through true hostilities or the fear they generated. He advocated giving them gifts of tools or clothes as a way for the government "to insinuate itself in the heart of a people without civility . . . to win their sympathies and reconcile them to civilized society."[53] The idea that "bárbaros" could be civilized and incorporated into society represents further continuity with colonial thinking.

But some white liberals now began to think that Christianization and sedentarization were not enough to civilize indigenous people. Prefect Carrasco, one of the principal proponents of liberal reform in Beni, wrote in 1851 of "the exceptional condition of these natives who, since their conquest, have not made one step toward civilization. Educated by the Jesuits with that community spirit which they still maintain, they profess a blind obedience consequent to that rigorous tutelage, more baneful than feudalism and even worse than the condition of the Helots [Spartan slaves]."[54] Carrasco thus saw no value in the Mojos' mission culture, and even rewrote Mojo history by claiming that they had been conquered (when in fact they had invited the Jesuits in).

Nineteenth-century Latin American liberals usually advocated education as a means to more fully incorporate Indians and strengthen the state, so we should not be surprised to find evidence of schools in Beni during the decades of liberal reform. The department of Santa Cruz appointed teachers in San Pedro, Exaltación, and Huacaraje in 1841. Fourteen Beni towns had elementary schools for boys in 1844, and the government paid the teachers' salaries. Prefect Borda introduced the Lancastrian method of using more advanced students to teach less-advanced ones in these schools, and opened a girls' school in Trinidad in 1846. A private school opened in the same town in 1850. Prefect Valverde opened girls' schools in Baures and Magdalena in 1849–51 but could not find teachers for them. Bishop Prado found schools at San Ramón, Magdalena, and Carmen in 1850, and Gibbon noted that the schools of Beni taught their lessons in Spanish. By 1868, all of the former mission towns had schools that taught the Indians to speak, and sometimes to read and write, in Spanish.[55]

Though most schools represent a continuation of earlier mission schools, liberals expressed a desire to modernize education and provide the department with secondary schools in the first years of reform. Mariano Arroyo, the general administrator of the Province of Mojos, lamented that the region did not have secondary schools with "scientific education" in 1844. Prefect Borja ordered the establishment of a secondary school for the mechanical arts in Exaltación in 1844, but Borda noted in 1846 that the order had never been complied with. Prefect Ibáñez took a special interest in education in 1848. First he gave the students of the Trinidad school sixty-two varas of Mojo-made cloth for clothing. Then he proposed the creation of a secondary school in San Pedro to the national government. Five boys from each cantón would be chosen to attend the school, and the state would provide for their transport, food, and clothing. The school was also to follow the Lancastrian method, so that teachers would instruct seven boys in the morning, each of whom would in turn teach ten others in the afternoon. Upon completing their education, these seventy students would return to their towns to open schools under the supervision of their priests.[56] This ambitious plan was never put into practice, and the proposed role of the church highlights how conservative educational reform was in Beni. The church retained administration of the schools since church and state remained partners in Bolivia, and education traditionally fell under church control. Whatever language might be used to describe it, and whatever plans the government had to modernize it, education in reform-era Beni can best be thought of as a continuation of colonial practice.

## Social Divisions

Although liberal reformers tried to dismantle much of Beni's mission culture, they also supported the continued existence of Indian cabildos and parcialidades as necessary for political, social, and economic order. The penalty of whipping, so much a part of the maintenance of order in the ex-missions, also continued. Prefect Aguilar outlawed whipping in early 1849, but by May he had restored it for thieves and for those who disobeyed their superiors, while complaining about the "ungrateful natives who don't recognize the liberalities of the government in making them citizens." Gibbon found twelve lashes a common punishment for "insubordination, drunkenness, or idleness" in 1852, but added that the authorities "punish whenever they deem it proper, with as many lashes as they please."[57]

The cabildos continued to function, and Gibbon described the cabildo of Trinidad in great detail after his extended visit. As the highest Indian officer, the corregidor was considered to be head of the cabildo by the carayanas; he reported directly to the prefect every Sunday. The cacique was the second highest officer and head of the cabildo in the strict sense; he received reports from lower officials and passed them to the corregidor every evening. Among these lower officials were the intendente and alférez, who oversaw portions of public business. Four *aguaciles* acted as constables, and eighteen *comisarios* fulfilled a variety of tasks, including carrying orders, keeping night watch, or performing duties in the prefecture such as waiting on the prefect's table. Two *policia* officers made sure that the boys of the town carried water, and sixteen *capitanes* commanded unpaid work gangs. A teniente de estancia or mayordomo de estancias oversaw the cattle herds on the pampas, while an alcalde was in charge of river transport, handling canoes and appointing crews of paddlers.[58]

Gibbon was impressed with the order and cleanliness of Trinidad, and attributed it to the regulations of the cabildo government. Nevertheless, Indian self-government under the cabildo still did not represent some kind of native utopia. Gibbon noted some rivalry between indigenous leaders when Cacique Mariano Cayuba said that he was better qualified to be corregidor than Frutos Nosa, who held that post.[59] Cabildo officers, often called *jueces* (judges), also used commoners to work their fields without pay.[60] In addition, cabildo elites in at least some towns began to marry whites and may have begun identifying more with the interests of their white relatives than with Indian commoners. Hipólito Ojopi, the cacique of Concepción de Baures, married the creole Cruceña María Manuela Vaca around 1841 (judging by the age of their eldest child), and their three mestizo sons became part of the town's elite.[61]

Prefect José Matías Carrasco enacted important regulations affecting the cabildos during his second term of office (1858–60). Each cantón was to have a cabildo, which was subordinated to the carayana municipal government. Among the cabildo members, the cacique was to communicate orders and reports between the corregidor and the other officers, attend to order and public works, and assign his people as laborers. The intendente was mainly responsible for the cleanliness and order of the town. Other officials, including the capitanes of the parcialidades, capitanes of the port, tenientes, alférez, alcaldes ordinarios, alcaldes provinciales, comisarios, fiscales, and alguaciles, all had some role in directing the different aspects of former mission life.[62]

The parcialidades still existed in the 1850s, although they were declining in importance by then. A padrón of 1854 lists the parcialidades of each Beni town in a way which suggests that the familia-pueblo ranking continued to exist (see table 2.1). Usually, Spanish-language industry names typical of the familia come first, followed by Mojo-language ethnic names (ending in the Arawakan plural –ono) characteristic of the pueblo. Cowboys (*baqueros*) vary in the position in which they are listed, but they usually appear either near the end of the industry names or the end of the entire list. Outlying cattle-raising settlements (estancias) also tend to be

**Table 2.1** Parcialidades of the four Mojo towns in 1854

| *Mojo town* | *Parcialidades (listed in the same order as in the padrón of 1854)* |
|---|---|
| San Javier | Músicos, Sacristanes, Carpinteros, Sastres, Herreros, Tejedores, Baqueros, Pojobocono, Moribocono, Mopiriono, Iyonos, Usuniono, Ortelanos |
| Trinidad | Músicos, Sacristanes, Carpinteros, Herreros, Tejedores, Sastres, Pintores, Canasteros, Baqueros del pueblo, Lozeros y Tejeros, Aperiono, Moyuneono, Cojucoreono, Chuchineono, Siyabocono, Achubocono, Tapimono, Estancia de San Miguel, Estancia de San Borja, Estancia de San Antonio, Estancia de Santa María, Estancia de Carmen, Estancia de San Lorenzo, Estancia de San Rafael |
| Loreto | Músicos, Sacristanes, Carpinteros, Tejedores, Herreros, Zapateros, Pintores, Enfermeros, Tejeros, Bonopiono, Suberiono, Munesono, Casaboyono, Chamuniono, Baqueros del Espíritu Santo, Baqueros de San Francisco, Baqueros de San Ignacio y el Carmen, Baqueros de San Rafael |
| San Ignacio | Músicos, Sacristanes, Carpinteros, Tejedores, Zapateros, Canasteros, Cosineros, Herreros, Carijiriono, Sunubono, Comobocono, Mures, Casabenono, Erisebocono, Tejeros, Baqueros, Ortelanos |

*Source*: República Boliviana, Departamento del Beni, *Padron jeneral de indijenas contribuyentes de la Provincia de Mojos, Año de 1854* (ANB RV Beni 13, T.1).

listed last. Where gardeners (*ortelanos*) exist as a parcialidad, they are listed below all other groups. Although it listed familia before pueblo, the padrón of 1854 reveals a blurring of these formerly important distinctions. The term *parcialidad*, previously applied only to ethnic divisions of the pueblo order, had now become generalized to mean any division of the population of a former mission town, whether based on occupation or original ethnic affiliation.[63]

By 1861, government records enumerated all of the Mojos and most of the other former mission Indians not by parcialidad, but rather by *manzana* (block), a system that continued until at least 1874. In this sense, manzana connotes a small unit of population and represents a Spanish translation of *îlot*, the smallest French census unit (based originally on an actual urban block). In the Mojo towns, all of the industry and ethnic parcialidad names were now replaced by liberal-patriotic manzana names honoring battles (Florida, Junín, Ayacucho, Ingavi), liberal political ideals (Libertad, Igualdad, Fraternidad, Independencia, Unión, Orden), or people (Franklin, Gutenberg, Jenner, Olañeta). In Trinidad, several estancias retained their old religious names, and in San Javier, *baquería* was used for cowboys. For the most part, these manzanas were merely old parcialidades

**Table 2.2** Manzanas of the four Mojo towns in 1861

| *Mojo town* | *Manzanas*[1] |
|---|---|
| Trinidad | Junín, Ayacucho, Libertad, Independencia, Ingavi, Dr. Olañeta, Dr. Franklin, Igualdad, Guttembert, Unión, Dr. Jenner, Estancia del Carmen, Estancia de San Lorenzo, Estancia de Rosario, Bolívar (forasteros only) |
| Loreto | Libertad de Músicos; Independencia de Sacristanes; Igualdad de Carpinteros, Tejedores, Herreros y Zapateros; Fraternidad de Pintores, Enfermeros y Tejeros; Florida; Ayacucho de Baqueros |
| San Ignacio | Independencia, Orden, Libertad, Igualdad, Ingabi, Ayacucho, Junín |
| San Javier | Junín, Ayacucho, Libertad, Independencia, Ingavi, Baquería |

*Source*: *Revisita of Beni*, 1861 (ANB TNC RV No. 10 1861).
[1] Names of manzanas follow spellings given in original documents.

**Table 2.3** Manzanas of the four Mojo towns in 1874

| *Mojo town* | *Manzanas*[1] |
|---|---|
| Trinidad | Junín, Ayacucho, Libertad, Independencia, Ingavi, Olañeta, Franklin, Igualdad, Gutember, Unión, Jenner, Estancia de Santa María, Estancia del Carmen, Estancia de San Lorenzo, Estancia de Rosario |
| Loreto | Libertad, Independencia, Igualdad, Fraternidad, Florida, Ayacucho |
| San Ignacio | Orden, Independencia, Livertad, Igualdad, Ingavi, Ayacucho, Junín |
| San Javier | Junín, Ayacucho, Libertad, Independencia, Ingavi, Baquería |

*Source*: *Matrícula jeneral del Beni para el Tribunal de Cuentas* (ANB RV Beni 1874 3).
[1] Names of manzanas follow spellings given in original documents.

renamed by liberals, and the manzanas of Baures and San Ramón retained industry and ethnic names. Even so, the match is not exact, as the number of manzanas in the Mojo towns in 1861 was lower than the number of parcialidades seven years earlier. The subdivisions of Loreto received hybrid names that illustrate how the transition from parcialidad to manzana could be made. The high-ranking Músicos and Sacristanes each received their own manzana, while the lower-ranking occupational parcialidades were combined several to a manzana. All five parcialidades with ethnic names became the manzana of Florida, while the four groups of cowboys were reduced to the manzana of Ayacucho de Baqueros. In Trinidad, carayana forasteros received their own manzana (Bolívar), but there were also forasteros distributed throughout seven other manzanas. The manzana of Bolívar no longer appears in government records from 1874, suggesting that carayanas were no longer paying tribute by then.[64]

The government showed more interest than the Indians in maintaining parcialidades as meaningful productive units dedicated to specific industries. The supreme order of April 20, 1847, specified that guilds (*gremios*) would be formed for each occupation and masters chosen to take possession of state-owned artisanal equipment. Prefect Ibañez established "tanners' guilds" in 1848, which suggests that no parcialidad was carrying out the tanning of hides then. In 1858–60, Prefect Carrasco ordered the corregidores to redivide the people into "parcialidades or guilds," as those institutions had fallen into disuse. Each parcialidad captain and his assistant were then to assign lands to their people to be worked communally, to provide crops of manioc, plantain, rice, squash, beans, peanuts, and sweet potatoes.[65] So despite attacks on communal property from 1842 to 1856, this official support allowed the members of at least some parcialidades to continue to hold collective title to land during and after this period, even into the 1920s.[66]

The parcialidades had lost most of their former significance in grouping people according to occupation by the time they were replaced by the manzanas. For example, the parcialidad of Sacristanes was not the same as those actually serving as sacristans. Since each town's parcialidad of Sacristanes normally became its second manzana in the 1861 revisita, we can determine whether serving sacristans came from the ex-parcialidad of Sacristanes when the records specify who held those offices. The two sacristans of San Ignacio and four of the sacristans of San Joaquín came from their second manzanas, and were thus Sacristanes. However, another two from San Joaquín came from its third manzana. The four sacristans of San Javier, the two of Exaltación, and the one of Magdalena were all from those towns' first manzanas representing their ex-parcialidades of Músicos. Two of Trinidad's sacristans (including the *sacristan mayor*) came from the tenth manzana (Unión), while the third came from the fourth manzana (Independencia); both manzanas were certainly descended from lower-ranking occupational parcialidades. Thus the number of men serving as sacristans was always quite small, and they were often not even from the parcialidad of the same name. Nevertheless, the sacristans held a high place in ex-mission Indian society, as evidenced by the outrage the people of Magdalena expressed to Bishop Prado at their priest's mistreatment of the sacristans. In addition, the revisita of 1861 lists church sacristans, singers, and musicians as reservados. They thus enjoyed the same exemption from tribute (impuesto personal) that the national government had granted to highland Indian church assistants in 1831.[67]

The revisita of 1861 also provides information on the parcialidad affiliations of indigenous leaders in several towns. In Loreto, Exaltación,

San Joaquín, San Ramón, and Reyes, the caciques came from the first manzana, which still bore the name Músicos in several towns. The corregidor of Loreto was also a Músico. However, the cacique of Magdalena was from the third manzana, the cacique of San Ignacio was from the seventh, and the cacique of Trinidad was from the eleventh. Trinidad's ex-corregidor Frutos Nosa was a member of the fourth manzana.[68] Cabildo leadership in 1861 thus remained with the highest-ranking parcialidad (the Músicos) in several towns, especially on the northern and western periphery, but came from lower-ranking parcialidades in other towns, mostly among the Mojos and Itonamas.

## Attempts to Improve Internal Communications

Bolivian leaders who dealt with Beni desired progress above all else during the period of liberal reforms, and the improvement of internal communication was one measure of progress. The supreme decree of November 18, 1842, which created the Department of Beni included among its objectives "to facilitate communications of those fertile regions with the interior departments, to open different ways of communication, and to multiply the colonies that are being established."[69] The development of Beni thus depended on occupying areas between the islands of settlement represented by the ex-missions and on improving communications beyond the simple navigation of rivers by indigenous boatmen.

In 1842, the authorities refounded San Borja, a mission town between San Ignacio and Reyes that had been abandoned. By 1844, they had also founded a new settlement at San José de la Sierra, to the southwest of San Ignacio. The bulk of the settlers of both places were local indigenous people. Limpias claims that Itonama married couples were the first to resettle San Borja, but the revisita of 1861 gives other details about the ethnic composition of San Borja. Twenty-two Borjano originarios (resettled from Reyes) formed the core of the population, but they were joined by several groups of forasteros: fifteen Movimas, thirteen Trinitarios, four Javieranos, four Ignacianos, and four carayana vecinos and government employees. San Borja also served as a base for the conversion and settlement of independent Chimane Indians. Despite many initial conversions, the Chimanes refused to settle permanently. A government official noted "the natural repugnance of the Chimanes to living with the Borjanos" and the fact that the Chimanes' own territories had better resources than San Borja as reasons for their reluctance.[70] Ignacianos played an important role in founding San José, opening a road there from San Ignacio, clearing land for its first crops, and contributing twelve carpenters to help build the first structures. At least two Mojo women (probably Ignacianas) were among the settlers of San José. They and all other women at the settlement spun cotton thread, while almost all the men worked agricultural plots.[71]

The early years of the department were marked by the feverish construction of "roads," which really meant charting new trails for the export of cattle. In 1844, Prefect Borja began two new roads. One ran from Atén to Mapiri, improving contact with La Paz. José Vicente Saavedra was to construct another, from Loreto to Carmen, in exchange for three hundred pesos' worth of state cattle. Governor Villegas had tried and failed to open the same route in the late 1830s as a way to avoid attacks by independent Indians on boats on the Iténez River. Also in 1844, Borja's subordinate José Agustín Palacios explored the lower Beni River and Lake Rogaguado and then inaugurated a road from Reyes to Irupana. By the time Borja left the office of prefect in April 1845, he could report the opening of other roads from Magdalena to Baures and San Simón, from Carmen to Guarayos, and from Reyes to Guanay. Borja believed that the road from Magdalena to Baures would be especially useful as a way to avoid attacks by independent Indians on trade on the Iténez River.[72]

Some of Borja's roads were actually still incomplete when he left office. Prefect Borda had to finish Saavedra's road from Loreto to Carmen later in 1845. He ordered twelve Loretanos and fifty Trinitarios to work on different sections under direction of their own cabildo officials, and all were to be rationed with meat from Mojo estancias. The Mojos also had to build a four-vara-high wooden cross every league to mark the way. The corregidor of Carmen was to provide workers to Saavedra to complete the other end of the road. Borda himself traveled the road from Carmen to Loreto in December 1845, but found it impractical and recommended rerouting part of it. The prefect later disputed Saavedra's right to collect his payment, pointing out that Saavedra did not complete the road.[73]

For the Mojo towns, the most important projects involved opening new overland routes toward Santa Cruz and Cochabamba. The prefecture gave the task of opening a road to Cochabamba to José Tudela, who failed, and then to Rafael Morales, who opened sixty-one leagues that could be covered in eight days in 1844, to a place near the foot of the Andes called Moleto. Twenty leagues from San Ignacio, the road passed through the newly repopulated San José de la Sierra. Late in 1845, Pablo Camino, the cacique of Loreto, blazed a new trail to the borders of the Department of Santa Cruz on his own initiative. Camino kept a detailed diary in Spanish of his trip, and described the land, its water features, and its vegetation with an interest and exactness befitting a native of the region. Borda praised Camino's work in a letter to the minister of the interior, saying it would shorten distances and provide "security to commercial speculations," presumably in cattle.[74]

Though the desire to open roads through the llanos was a key part of liberal plans for progress, it was not unique to the mid-nineteenth century. Government officials and private individuals frequently proposed and opened overland trails from the colonial period into the twentieth century, often along the same routes as if no overland connection had ever existed before. In 1849, for example, Mojo laborers began opening a road from San José de la Sierra to Cochabamba, along the same trail used by Morales.[75] The climate of the llanos ensured that such paths would indeed frequently disappear, either through the effects of floods or the exuberant growth of vegetation.

## The Opening of Trade with Pará

Another aspect of liberal reform was the desire to increase foreign trade. Classical liberal theory advocated a lowering of international trade barriers so that the economy could function more efficiently according to its own supposedly rational laws. In 1837, as free trade was gaining momentum in Europe, President Santa Cruz promulgated a decree of free trade for Bolivia, which did not immediately affect Mojos. During Santa Cruz's short-lived Peru-Bolivian Confederation (1836–39), Bolivians had free access to the Pacific port of Arica in Peru. Though the Madeira River offered a possible trade route through Brazil to the Atlantic, the middle Amazon was still recovering from the Cabanagem rebellion and remained insecure until the last Cabano rebels laid down their arms in 1840. It thus fell to José Ballivián to promote trade between Beni and Brazil's Amazonian province of Pará. Ballivián's defeat of a Peruvian invasion in 1841 marked the final separation of Bolivia and Peru, and left Arica in Peruvian hands. Bolivia was forced to develop the less-convenient Cobija as its Pacific port, and the search for an Atlantic trade outlet then assumed greater importance.

As early as 1840, the Briton Alexander Campbell made a river voyage between Mojos and Pará. In 1841, the Bolivians Tadeo Goariti and Nicolás Román traveled from Exaltación to Pará. In 1844, the Beni government took the initiative by organizing a "Flotilla Beniana" to trade with Pará, delegating the task to León Hurtado, corregidor of San Ramón. The flotilla returned the next year with a small amount of iron and steel, insufficient to meet the needs of Beni. Hurtado nevertheless informed the prefect that trade with Pará was practicable, and that Brazilians had exaggerated the difficulty of the cachuelas to keep the Madeira to themselves.[76]

In early 1845, the prefecture contracted with Joseph William Fields (recorded as José Guillermo Campos), a carpenter from New York resident at Exaltación, to build boats with Indian labor for the Pará trade. In January 1846, the prefect asked the national government to send tools from La Paz or Cochabamba in order to prepare the second expedition of the Flotilla Beniana to Pará. After a series of problems, Fields informed the prefect in November 1846 that the trading expedition would be postponed, and that the Indians had returned to their respective towns.[77] Even so, José Agustín Palacios led an exploratory expedition of five boats and seventy-one Indian crewmen out of Exaltación in October 1846. Palacios explored the cachuela area and descended to the middle Madeira, returning to advise Prefect Borda that a trade route could be opened if the government would dredge, straighten, or build roads around the cachuelas and introduce small steamboats. But shortly thereafter, Borda wrote that Brazil had expelled some Bolivian traders from Pará, that Bolivian traders had no guarantees in Brazil, and that the Madeira only offered "barbarous tribes and the small [Brazilian] village of Borba." Thus when the Brazilian Joaquim Teodoro Tocantins approached Prefect Moreno in 1848, offering to establish a dockyard and build boats with foreign carpenters in exchange for thirty state-owned boats, Indian labor, title to forest land for timber, and a contract to provide river transport, nothing came of his proposal.[78]

We know little of the early Beni trading expeditions, but it is safe to assume that their crews were composed of former mission Indians. At least two Loretanos traveled with the Flotilla Beniana. While away from home, crew members behaved very independently, often in ways harmful to their carayana bosses or *patrones*. Hurtado complained that his crewmen had cheated him while trading in Brazil. More seriously, crewmen could easily kill their patrones while on isolated parts of the rivers. Goariti and Román were killed while returning to Exaltación from Brazil in 1841, and three natives of Carmen killed Campbell in 1845.[79]

The flurry of official interest in river trade subsided during the protectionist administrations of Belzu and Córdova, between 1848 and 1857. Prefect Valverde (1849–51) did send another flotilla with tallow and hides to Pará, but it only returned with spices and tea, and had to fly the Brazilian flag as Brazil had prohibited Bolivian boats in its territory. Gibbon's voyage of 1852 took place during this lull in trade and was possible only because he accompanied a returning Brazilian trade expedition.[80] As long as the production of temporalidades could be exploited by carayana and cabildo elites, they had little reason to look for profit in the Pará trade, especially while such trade was illegal. The voyage from Beni to Pará had proved to be dangerous while producing little profit. Nevertheless, the trade expeditions of the 1840s laid the groundwork for later voyages, which began again in the 1860s.

Despite the years of reform, the culture of the former mission Indians of Beni in the mid-nineteenth century was still firmly based in the colonial experience and had changed little since then. The settled population of the department still lived in or near the old Jesuit mission towns and was still predominantly indigenous, with only a small number of carayanas. Though the Indians traveled the rivers of the region between their own settlements and en route to Santa Cruz and Cochabamba, they disliked long river voyages. Gibbon noted that the typical Mojo "does not travel abroad himself, but remains in his own district."[81] Their world was still very isolated. The former mission towns with their fields and pastures formed an island surrounded on all sides by unincorporated indigenous peoples, including the Sirionós to the east, the Chimanes to the west, the Yuracarés astride the river route to Cochabamba, and the Caripunas around the Madeira-Mamoré rapids. In 1860, there was still little reason for outsiders to come to Beni, and little reason for the natives to leave. This situation was soon to change, as we shall see in the next chapter.

## Conclusion

In their desire to modernize society, liberal reformers hoped to stimulate production by making the Indians private property holders and relieving them of the burden of colonial-style forced labor. Yet in their push for rapid development, liberals demanded more labor than ever in order to reform the urban plan and open new routes of trade and communication. Liberals quickly discovered the practical limits of their ideology and fell back on colonial models. A tax originally levied on men of all races soon became an Indian tribute like those of the highlands. The new urban plan mimicked colonial highland towns, education showed little change from what the missions had offered, and church and state remained partners. Citizen became a synonym for vecino, and only the most important Indians were referred to as citizens by whites. The reform period ended with a new imposition of unpaid labor in the service of civil and spiritual authorities, a quick return to communal labor, and a reaffirmation of cabildos and parcialidades as organizers of labor. The end results of reform—including new taxes, the opening of land to private ownership, and the termination of mission-style redistribution and of the state's reciprocal obligations—were mostly prejudicial to the Indians.

Yet the same sources that document government attempts to reform Indian society also give us of a glimpse of Indians as actors. Indian leaders embraced the concept of citizenship, especially in Trinidad. The Mojos and other ex-mission Indians participated in carayana political squabbles, and occasionally killed white officials and patrones. The decline of some elements of the mission legacy, such as the parcialidades and their production, stems from the Mojos' conscious decision to discard institutions that no longer served their interest once the state had curtailed the benefits of mission-style redistribution. As artisan production began to decline, the Indians blurred the distinction between the colonial categories of familia and pueblo by calling all subdivisions parcialidades and supporting leaders from various parcialidades. And as they came to realize that river transport was the only work that provided a wage, it is perhaps not surprising that the ex-mission Indians would turn more and more to this occupation in the 1860s.

CHAPTER THREE

# A Country Vulcanized

## A New Economy

Life along the Bolivian tributaries of the Amazon changed dramatically with demands of the international market for tropical forest products, beginning with a limited trade in cinchona bark, the source of quinine, in the 1840s and 1850s. At the same time, Charles Goodyear's 1839 invention of the vulcanization process (stabilizing rubber by heating it with sulfur) created a seemingly unquenchable demand for rubber for the numerous flexible components of industrial machinery. The subsequent history of the rubber industry, with its constantly increasing profits, is well known. Less well known is the local impact of the rubber industry on the many communities and regions that make up the Amazon. One Bolivian rubber patrón-turned-prefect summed up that impact in the language of the industry by saying that Beni had become "a country vulcanized."[1] The transformation implied by this phrase began in the 1860s, when Bolivian patrones and workers returned to the Madeira River to trade, and were soon drawn into rubber commerce as well. As river trade and rubber production grew, ex-mission Indians from Beni increasingly left their towns to serve as rubber gatherers and boatmen on the Madeira and other northern rivers. Because accounts from the rubber zone usually did not specify the ethnicity of Bolivian Indian workers, I will use the term "Beni Indians" to refer to Mojos, Canichanas, Movimas, Cayubabas, Baures, and Itonamas.

The presidency of José María Linares (1857–61) signaled the return of free trade to Bolivia. Linares lowered tariffs protecting the native cloth industry, opening Bolivia to mass-produced British textiles, and improved the climate for trade by abolishing the *peso feble*, a debased currency that had been used since 1829. Travelers began to descend from Bolivia to the Amazon River by the time Fray Samuel Luciani made such a journey in 1858. By the early 1860s, three or four Bolivian boats descended the Madeira every year, and Brazilians had come to recognize Beni Indians by their distinctive camijetas by 1862. By 1864–65, about seventy Bolivian boats per year were descending to Serpa (now Itacoatiara) near the mouth of the Madeira, carrying jerky, cacao, cattle, deer hides, tallow, tobacco, cigars, and sugar. Brazilian officials estimated that the value of Bolivian products grew from less than 20,000 milréis in 1861 to 120,000 milréis in 1865. The latter figure may be an exaggeration, since other Brazilian statistics show the value of Bolivian exports and imports through the Province of Amazonas as rising from 64,004 milréis in 1865, to 84,750 milréis in 1866 and 85,734 milréis in 1867. At any rate, the river trade grew substantially, and when Brazil opened the Amazon to unrestricted international traffic in 1867, traders from Beni had access to more goods than ever before.[2]

Brazilian accounts give some idea of the movements of Bolivian merchants and their Indian crews along the middle Amazon in the 1860s and 1870s. According to Francisco Bernardino de Souza, every year Bolivian boats descended to the ports of Serpa or Vila Bela da Imperatriz (now Parintins). Once on the middle Amazon, the merchants had two options. They could exchange their Amazonian products for foreign merchandise at one of the two ports, and then direct their boats behind the Island of Tupinambaranas to the town of Maués. There they would pick up cargoes of guaraná, a substance prepared from the fruit of a tree of the same name, which was much in demand in Bolivia and Mato Grosso to make a refreshing, highly caffeinated drink. From Maués the merchants and their crews could continue on to the lower Madeira, and begin the voyage of three months or more back to Bolivia. The other alternative was for merchants to leave their boats in Serpa and descend to Belém with their goods on a steamboat. In Belém, they would sell their products and buy foreign goods before returning to Serpa by steamboat, and to Bolivia with their boats and crews. The Brazilian lawyer and politician Aureliano Cândido Tavares Bastos, who traveled on the Amazon in 1865, noted that Bolivians sold cacao, tobacco, hides, and other products in Belém. In that port, they bought iron bars, axes, agricultural implements, gunpowder, guns, liquor, and baubles to bring back to Bolivia. The same author observed that Bolivian crews would leave Serpa for Manaus while their patrones were in Belém. There, the Indians labored on public works in the rapidly growing city while waiting for their patrones to return. Outside Serpa, other Bolivian workers tended a plantation of eighteen thousand cotton plants established by a Bolivian patrón by 1864 to profit from high prices during the American Civil War.[3]

Foreign travelers also noted the presence of Bolivian Indians on the middle and lower Amazon. In 1865, Louis and Elizabeth Cabot Cary Agassiz encountered one Beni Indian among several Brazilian boatman on the Amazon between Santarém and Manaus. In the same year, the Agassizes traveled to Maués with the president of Pará. When they reached the beach of Maués, they found "a large party of Bolivian Indians, who had built their camp-fires on its sands." When the Indians found out that there were white visitors in the town, they crowded outside the door of the president's lodging to sell their fine cotton camijetas. These crews were returning to Bolivia from Serpa or Vila Bela da Imperatriz, and passing through Maués to trade for guaraná. James Orton described the clothes of the Indians he saw at Serpa in 1867 as "a long gown of bark-cloth reaching to the knees." Although he did not identify them as Bolivian, they undoubtedly came from the ex-missions of Beni, where men often wore bark cloth shirts instead of cotton camijetas while working.[4]

## The Madeira Barracas

Because of this lively river trade, Bolivians were well acquainted with the Madeira and the middle Amazon by the time they began to exploit rubber. In 1864, José Santos Mercado founded the first Bolivian rubber extraction settlement, or *barraca*, in the cachuela zone at the confluence of the Yata and Mamoré rivers, and began exporting rubber in 1865. In 1866, four brothers from Santa Cruz named José Manuel, Antonio, Querubín, and Antenor Vásquez opened a second zone of extraction by constructing two barracas on a tributary of the Beni River. Shortly thereafter, Santos Mercado moved his operations downstream from the Mamoré to the Madeira River.[5] Bolivia had inherited the Spanish claim, from the 1777 Treaty of San Ildefonso with Portugal, to much of the left bank of the Madeira, and Bolivian *siringueros* (rubber tappers) soon moved into that area.

By the 1867 treaty between Bolivia and Brazil, their boundary was moved a considerable distance to the south, to a line extending northwest from the Beni-Mamoré confluence toward the source of the Javari River. Despite historian Valerie Fifer's statement that "the small numbers of Bolivians involved fell back, choosing for the most part to abandon their claims along the Madeira River rather than be stranded in Brazilian territory," Bolivian rubber patrones and their Indian workers remained on the

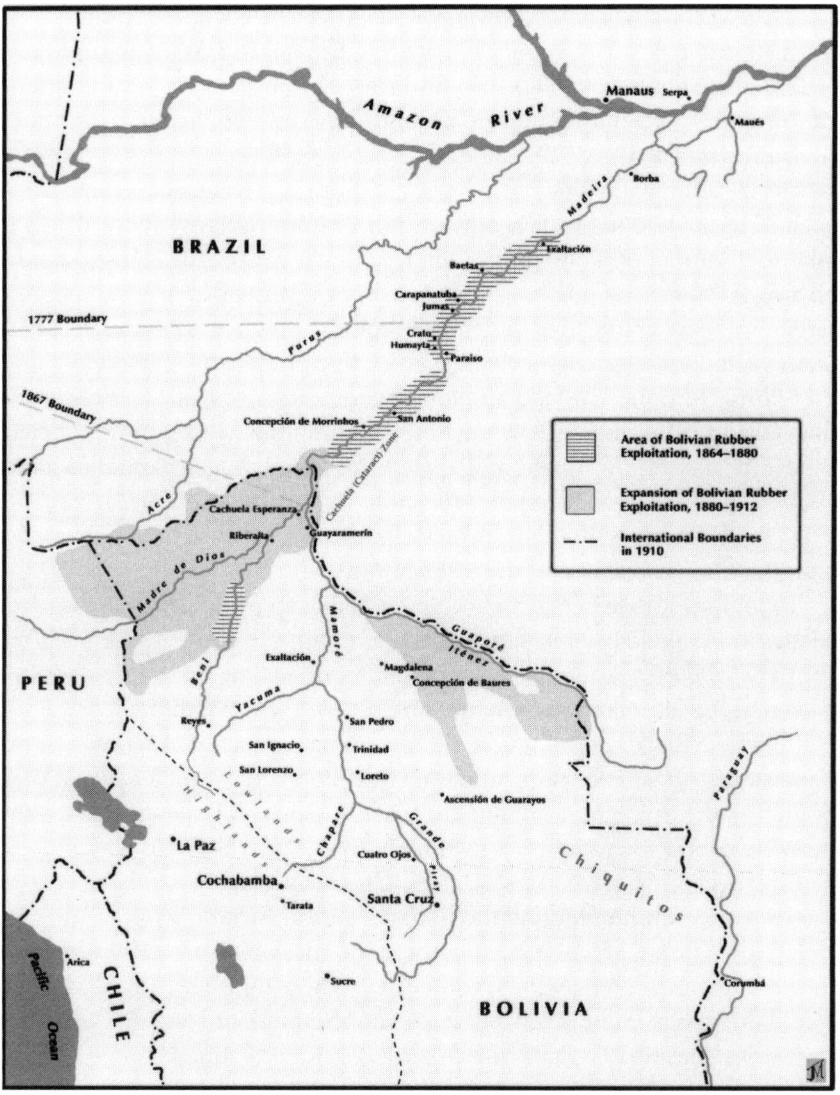


*Map 3.1.* The Bolivian rubber boom. Map by James Martin.

Madeira for many years afterward.[6] In fact, they exploited the right bank, which had been recognized as Brazilian since the eighteenth century, as often as the formerly Bolivian left bank.

Travelers' accounts that provide a description of the extent of Bolivian rubber exploitation on the Brazilian Madeira include those of the German engineers Josef Keller and his son Franz (1868), the Italian missionary Fray

Jesualdo Maccheti (1869), the English engineer Edward Mathews (1874), the English travelers Charles Barrington Brown and William Lidstone (1874), the Brazilian traveler João Severino da Fonseca (1877), the American engineer Neville Craig (1878), and the Austrian-Brazilian engineer Júlio Pinkas (1883).

Franz Keller wrote that there were "ten or twelve Bolivian Seringueiros, each of them working with twenty or thirty Mojos Indians" above the Brazilian settlement of Crato on the Madeira River (in Brazil, "Mojos" could be any of the former mission Indians of Beni). At Crato itself, he encountered Bolivian cattle that had been brought down the Madeira. Maccheti visited four Bolivian rubber patrones just below the cachuela zone, including Nemesio Hurtado and the Cruceño Pastor Oyola, and observed that almost all siringueros were Bolivian Indians. Maccheti also noted the Bolivian cattle of Crato and visited the house, sugar plantation, and rubber forests of Ignacio Araúz, the Bolivian vice consul for the region. This, the Bolivian settlement furthest downstream, was named Exaltación after the Cayubaba town of the same name.[7]

Mathews estimated that "a thousand Bolivian peons" lived along the Madeira. He called Exaltación "one of the largest rubber settlements on the river" and met boatloads of Baures and Cruceño mestizos bound for the Madeira, the latter for the settlement of Baetas. Brown and Lidstone found "numerous villages and fasendas . . . occupied principally by Bolivians of Spanish or Indian extraction" and said that Bolivians formed "a large proportion of the inhabitants" along the Madeira River. A Bolivian consul and his wife lived at the mouth of the Jamary River, and a Bolivian patrón lived with his Indian workers at Exaltación. One year later (in 1875), agents of the British government found the Bolivian patrón Ricardo Chávez and two hundred "Moxos" at Carapanatuba (a name meaning "many mosquitoes" in the Tupían *lingua geral*). In the same year, a number of Bolivians on the Madeira donated money to the Manaus hospital, or Casa de Caridade, including Rómulo Ojopi, the mestizo son of Baure cacique Hipólito Ojopi.[8]

Fonseca met the Bolivian Angel Chávez and his wife near the cachuela of Girão, bringing fifty male and thirty female Indian workers to the Madeira from Trinidad. Later, he visited Chávez's barraca of Esperanza between the cachuelas of Caldeirão do Inferno and Morrinhos, and saw Chávez's one-story, bamboo-and-wood house among the dwellings of his 150 workers. Below Caldeirão do Inferno, Fonseca encountered the one-story, zinc-roofed house of Ignacio Araúz, surrounded by workers' shacks. On the Madeira below the last cachuelas, he found Santos Mercado living at the settlement of Paraíso.[9]

Craig found 180 men and 90 women at the Bolivian settlement of Jumas. He encountered a *cachaça* (cane liquor) distillery there, along with sugarcane, bananas, plantains, manioc, Brazil nuts, and a ricelike cereal. At Paraíso, Craig visited the home and the banana, plantain, and coffee plantations of Santos Mercado, and found Portuguese and Spanish Jews and "a large body of Bolivian Indians," including women. Above the cachuela of Morrinhos was the settlement of Concepción de Morrinhos, where Pastor Oyola lived with fifty to sixty servants. At the cachuela of Três Irmãos resided Ignacio Araúz and thirty other people, including Bolivian rubber gatherers and house servants and four Caripuna Indians. Araúz had another establishment at the rapids of Caldeirão do Inferno, which included an ox-powered sugar mill. Craig also mentioned Santos Mercado's house in the new settlement of San Antonio (Santo Antônio, near the modern Pôrto Velho), and Araúz's several other plantations on the Madeira and on the Amazon itself.[10]

Araúz had died by the time of Pinkas's visit. Nevertheless, Pinkas found the upper Madeira in 1883 still populated by "exclusively Bolivian citizens . . . accompanied by domesticated Bolivian Indians from the provinces of Mojos, Trinidad, or Exaltación." At the barraca of Carmen at the mouth of the Jaciparaná River, Pinkas encountered the Bolivian patrón Tristán Roca and his sixty male and forty female Beni Indians servants. Nearby was the establishment of another Bolivian, Enrique Viscarra. Between the juncture of the Mamoré and Beni rivers and the uppermost cachuela, that of Guayaramerín, there were eight rubber establishments with 265 inhabitants. Seven of these establishments and 215 of the inhabitants were Bolivian. An anonymous member of Pinkas's commission wrote that he first encountered Bolivian Indian workers (Trinitarios and Cayubabas) at Jumas while ascending the Madeira. He also observed that Santos Mercado, now consul-general for Bolivia, still lived at Paraíso with his wife. At the foot of the cachuelas, this traveler found about 300 "Bolivians and Indians" waiting for their patrones to come back from Manaus by steamboat, so that they could all return to Bolivia.[11]

## Upper River Barracas

Some Bolivians found rubber along the upper Iténez River, which the Bolivian writer Fabián Vaca Chávez later called "the antechamber of the empire of rubber." More importantly, rubber exploitation on the middle Beni River increased greatly. There, many cinchona bark gatherers shifted to rubber gathering as the price of cinchona fell while that of rubber rose. The Cruceños Pablo Salinas, Francisco Cárdenas, and Rafaela Pardo first brought rubber out of Beni River forests in 1869 and 1870. In 1870, they were followed by several other entrepreneurs, including the Cruceño Angel Arteaga, who founded the barraca of Cayuvaba with his Cayubaba crew on the Madidi River, a tributary of the Beni. The Trinidad-born Dr. Antonio Vaca Díez began collecting rubber on the Beni in 1876, with Cruceño mestizo workers. Antenor Vásquez became an important rubber patrón on the Beni River around the same time. By 1880, eleven barracas existed, with a total of 225 rubber workers. In 1881–82, Spanish missionary Fray Nicolás Armentia noted many barracas along the Beni and Madidi rivers.[12]

The early rubber patrones of the Beni River exported their rubber by a circuitous route upstream to Reyes, overland by oxcart to the Yacuma River, and then down the Yacuma, Mamoré, and Madeira rivers. Though Bolivians knew that the Beni flowed into the Madeira, it took the 1880 voyage of the American doctor Edwin Heath, who descended the Beni River to its mouth and then returned via the Mamoré and Yacuma rivers, to convince the Beni River patrones that a more direct route was practible.[13]

The Beni River soon became an important trade route. It was shorter, had only one significant rapid at Cachuela Esperanza, and its mouth was downstream from several of the Madeira-Mamoré cachuelas. In the 1880s, the Bolivian rubber industry experienced its most rapid growth as new areas were discovered and exploited. The Cruceño rubber trader Nicolás Suárez founded his base of operations at Cachuela Esperanza in 1881, and Antonio Vaca Díez and Antenor Vásquez soon followed him to the lower Beni. By the end of 1882, Vaca Díez had eight barracas on the lower Beni, with 181 rubber workers. Vásquez settled on a high riverbank that later became the city of Riberalta, and a French firm, Braillard, Clausen and Company, also moved to this site in 1884. In 1883, Pinkas listed 2,050 rubber workers on the Beni River, 50 on the Río Madidi, and 520 on the Río Madre de Dios. The Bolivian occupation of the lower Beni led to even greater rubber exploitation, as Bolivian and foreign rubber workers explored its many tributaries, including the Madre de Dios, Orthon, Tahuamanu, and Manuripi rivers, and found rubber trees everywhere. The native peoples of this area, the Pacahuara, Toromona, Araona, and Esse Ejja Indians, had previously had little or no contact with Bolivian or Brazilian society and were decimated by European diseases and firearms. Bolivian expansion in this region was halted only when Bolivians crossed into the Acre

River region and encountered Brazilian rubber tappers advancing from the Purus River.[14]

Born in Santa Cruz and raised in Trinidad, Nicolás Suárez and his three brothers became the most powerful rubber patrones and founded a firm eventually known as Suárez Hermanos. Francisco, the eldest, started the family business by providing supplies and credit to cinchona and rubber gatherers. After spending a year in London and serving as Bolivian consul-general, he returned to Trinidad and held the offices of municipal council president and interim prefect in 1872. He returned to London in 1877, established a European branch of the firm, and remained titular head of its Trinidad branch until 1890. Rómulo, the second brother, set up Suárez agencies in Brazil at Belém, Manaus, and San Antonio, all key points on the rubber trade route. He returned to Trinidad, acquired land for ranches, and settled north of town at an isla the Mojos called Monovi, and later Loma Ayacucho, indicating that it had belonged to the manzana of Ayacucho (the ex-parcialidad of Sacristanes). Under Suárez ownership, this valuable property became known as simply La Loma, and finally Loma Suárez. Another Suárez acquisition south of Trinidad likewise resulted in the Laguna Socoreno being renamed the Laguna Suárez. Rómulo sent jerked beef from these ranches, and sugar and rice from Santa Cruz, north to provision Suárez workers in the rubber zone. The third brother, Gregorio, was stationed at the Madeira-Mamoré rapids with five hundred Indian workers to supervise transport through that zone, while Nicolás, the youngest, remained at Cachuela Esperanza. The four brothers were thus able to coordinate all aspects of their rubber trade, from source to final market. Though the Suárez firm initially only traded rubber bought from others, it would later own the majority of Bolivian barracas, acquiring many from their indebted patrones.[15]

## The Impact on Indians: Introduction

This rubber boom had a significant impact on Indians throughout the Amazon basin. United States commercial attaché to Brazil W. L. Schurz summed up his impression of this impact in 1925:

> The more amenable tribes tend to lose their identity through miscegenation or other peaceful processes. . . . The less pacific or more helpless peoples are gradually being exterminated by intertribal strife or in clashes with the rubber gatherers, and disease has also taken a heavy toll

> among some of these folk. During the time of the great rubber boom the Indians were often relentlessly hunted, either as labor material or as irreconcilable opponents of the invaders of their tribal lands. This occurred on a large scale in the valley of the Javary, in the Acre country, and in the Itenez basin in Bolivia, where the population is said to have been decimated in the search for laborers for the *gomales* or rubber forests of the Beni.[16]

As this description suggests, the Amazon rubber boom is generally conceived of as a period when the region's Native American peoples were thoroughly brutalized. These Indians are usually considered to have been largely free of outside influence or control, before being virtually enslaved as rubber workers or else exterminated in bloody conflicts with the rubber patrones who invaded their territories. This picture is most true of the Putumayo region of northwestern Amazonia, which came to world attention when the American Walter Hardenburg published an account of the brutality of its rubber boom in 1909. Recent scholarly treatments by Michael Taussig and Michael Stanfield have continued to focus attention on the Putumayo.[17] In Bolivia, native peoples along the lower Beni River and its tributaries were indeed decimated by the rubber boom.

But Amazonia is a huge region with many contrasting histories, such as that of the Mundurucú of central Brazil, who voluntarily gathered rubber for the market.[18] And rubber workers included not only independent Indians with little previous outside contact, but also ex-mission Indians such as the Napo Quichua of eastern Ecuador. How did the Mojos and the other former mission Indians of Beni fare during the rubber boom? Many secondary and some primary sources paint the following picture: after the discovery of rubber, whites forced Beni Indians to settle in the rubber zones or ply the rivers for the rubber trade. Most Indians never came home again, and a great many died from abuse, tropical fevers, drowning in the cachuelas, or conflicts with independent Indians. The Bolivian rubber boom resulted in a massive depopulation of the former mission towns of Beni and the destruction of their mission culture.[19]

The above outline depicts the former mission Indians of Beni as helpless victims who lost both their lives and culture during the rubber boom, and largely denies their capacity for agency. This view fits in well with the existing historiography of the Amazon rubber boom, which often portrays the rapid and brutal destruction of previously independent Indian groups. But as the Mojos and most of their neighbors had been incorporated into a wider Latin American society for more than 150 years, we might expect their experience of the rubber boom to be different. The questions we have to ask to understand this experience include the following: What was the quality of life of former mission Indians involved in the rubber boom? What were the hardships and dangers of their work, and how were they treated by whites? How were Bolivian Indians recruited to work in the rubber trade? Were they forced? What were their relations with independent Indians like? How badly was Beni depopulated and was the rubber boom the cause? And what was life like in the former mission towns during this period?

## The Lives of Boatmen

With regard to the quality of life of Beni Indian workers in the rubber industry, we must take into account the difference between boatmen and rubber gatherers. The llanos natives had traveled the rivers of their region before the arrival of the Spanish, and on the eve of the rubber boom Indian crews were still indispensable for travel between the former mission towns, or to Santa Cruz or Cochabamba. In the early Bolivian rubber boom, Indian crews were even more essential. Steam navigation began on the Amazon River between Belém and Manaus in 1853, but was slow in coming to its southern tributaries. In 1869, Maccheti found that steamboats could ascend the Madeira River to varying distances based on the amount of water in the river. Several small, private steamboats operated on the Madeira in 1871, followed by regular steamboat service and the opening of the Madeira to foreign trade in 1873, and monthly service by 1877. Despite some earlier attempts, steamboats did not begin to ply the Mamoré and Beni rivers regularly until the 1890s.[20]

Merchants engaged in river commerce between Beni and Brazil before Bolivians began to exploit rubber, and even after 1864, these merchants' cargoes included many products other than rubber. Franz Keller estimated that sixty to seventy Bolivian boats annually descended the Madeira River with hides and tallow, and returned to Bolivia with European and North American manufactured goods. The boats that Maccheti traveled to Brazil with were filled with cinchona bark, hides, and tallow. Cinchona merchants only started exporting to the Atlantic rather than to the Pacific in 1868, and while the Madeira route was cheaper, the cinchona boom was already coming to an end. Though Mathews called rubber "the only trade of consequence on the Madeira," he noted that Brazil nuts and Sapucaya nuts were gathered and exported, and that traders carried provisions and guaraná up the Madeira. But European manufactured goods could also reach Trinidad from the Paraguay River port of Corumbá or from the Pacific coast, and sell for equal or lower prices than most goods brought up the Madeira. Hence the principal value of the Madeira was for the access it provided to rubber.[21]

Franz Keller described several boats of differing size used on the Madeira and gave their Brazilian names. A *coberta* was "a schooner with a sort of wooden awning on deck," a *batelão* was "a broad sloop with an arched covering of palm leaves," an *igarité* was "a smaller half-covered boat for rowing and sailing," and a *montaria* was a canoe. The shelter formed by the roof at the stern of most of these boats was called a *camarote*. In Bolivian Spanish, a batelão was called a *batelón*, and an igarité was a *garitea*. The three larger types of boat could only be afforded by merchants, as Keller says that a ten-ton igarité cost three hundred milréis, or forty pounds. Mathews, who did not differentiate between batelones and gariteas, said that they were eight to ten yards long and four to five across, drew two to four feet of water, carried three to six tons, and had crews of at least ten plus a captain. Craig's boat was twenty-five feet long, eight feet wide, with a draught of two feet, and had ten Indian paddlers and a captain. A ton or more of cargo was piled in the center of the boat, between the paddlers; it was because the cargo took so much space that oars could not be used. The boat of the American colonel George Church, who headed the first attempt to build the Madeira-Mamoré Railway in the 1870s, weighed some 3,225 pounds unloaded, and was twenty-four feet long, six feet wide, and two feet deep. Fifer provides similar dimensions for a batelón: thirty feet long and eight feet across.[22]

An expedition frequently included boats of different sizes and crews from different ethnic groups. Franz Keller's expedition consisted of eighty Mojo and Canichana paddlers, with sixteen crewmen in the largest boat and three in the smallest. Maccheti's expedition included more than 170 Indians, including Canichanas, Trinitarios, and Cayubabas, in thirteen boats, of which the largest was a batelón with sixteen crewmen. Mathews's group included seven boats of different sizes. The merchant Miguel Cuellas owned the three largest boats, two of which had Baure crews, and one of which had an Itonama crew. Mathews's boat had a multiethnic crew, including Cayubabas and Canichanas, and its members did not work as well together as did those of Cuellas's boats. Church's boat also had a multiethnic crew consisting of six Canichanas and six men from other Beni Indian groups. Fonseca's captain recruited a crew in San Joaquín composed of twelve Baures, one Cayubaba, and one Itonama.[23]

These Beni Indian boatmen paddled the Mamoré, Iténez, and Beni rivers and their tributaries, and the Madeira and Amazon rivers as far as Serpa, Manaus, and Belém. At least one boatman even traveled to the Orinoco River. It was hard work and, of course, was harder going upstream than downstream. The most grueling part of an expedition was getting around the eighteen Madeira-Mamoré rapids and falls, a task that ended only with the completion of the Madeira-Mamoré Railway on the Brazilian shore in 1912. Such a trip through the cachuelas could take three weeks going downstream, and nine weeks going upstream. At the larger falls of Teotônio, Girão, Ribeirão, and Bananeiras, the freight had to be unloaded and the boats dragged up to a thousand yards on log rollers. Boats had to travel in groups of at least three, because the crew of a single boat could not provide enough muscle power to haul their boat alone. Such a portage could take two or three days. At smaller rapids, boats could be left in the water and hauled upstream by rope from the riverbank.[24]

Even where there were no rapids, a boatman's workday was long and hard. Franz Keller's crew would get up at dawn, paddle for three to four hours, and then stop for breakfast and a two hours' rest. They would paddle two or three more hours, rest again, and then paddle until the evening's

*Figure 3.1.* Portaging around the Teotônio Falls on the Madeira in the 1860s. From Franz Keller-Leuzinger, *The Amazon and Madeira Rivers* (Philadelphia: J. B. Lippincott, 1875).

stop and dinner. Paddlers could get much less rest when their patrones thought it necessary to travel rapidly. When Mathews's expedition traveled through an area where the merchants feared attack by Moré Indians, the crews paddled upstream for up to eighteen hours per day and went for up to eleven hours without a break. On one day, they paddled from midnight to eight or nine in the morning, stopped for breakfast, and then paddled again until dinner at five or six in the evening. Church estimated that his paddlers made fifty-four strokes per minute on average, and forty-eight per minute going upstream, for up to ten hours a day. Their paddles were over four feet long and made of a hard, heavy wood. Church calculated that his crew's average speed on still water would have been about three and a half miles per hour.[25]

The boatmen's food included both supplies they brought along and other edibles they acquired en route. Franz Keller's crews ate a mush of corn or manioc flour, fresh or dried fish, jerky, and caiman. Maccheti's crew ate flour, caimans, and a lizard, and stopped to fish and to gather *almendras* (Brazil nuts). Mathews's crew ate *chupe*, a thick soup of jerky, rice, and manioc flour. They also refreshed themselves regularly with "sheebee" (the Brazilian *xibé*, adopted into Beni Spanish as *chivé*), a handful of manioc flour mixed with water that was drunk out of a "tortuma" (a *tutuma*, or gourd vessel). Only higher-ranking crewmen added sugar to their chivé. At various points along the way they acquired fresh plantains, corn, pumpkins, birds, monkeys, lizards, caimans, capybaras, otters, and fish. Franz Keller, Maccheti, and Mathews also obtained fresh produce through trade with independent indigenous peoples.[26]

Franz Keller and Maccheti both mention Indians going ashore for gulls' and turtles' eggs. Turtles were easy to catch and their meat was relished. Newly hatched turtles were considered a delicacy. The Praia de Tamanduá, a turtle nesting–ground on the Madeira River below the cachuelas, provided large numbers of eggs. Hundreds of fishermen and rubber workers collected these eggs to obtain "turtle butter," which was used as lamp oil. Brazilians regularly gathered eggs there at least as early as 1852, before Bolivians had settled on the Madeira. Gibbon and Franz Keller both estimated that such exploitation destroyed millions of eggs every year.[27]

Both Mathews and the Kellers noted that patrones would not carry enough food to last through the whole voyage. The merchants leading Mathews's expedition expected to catch enough game to make up for the provisions they lacked. The Kellers wrote that Bolivian crewmen would constantly rob the plantations of food crops at rubber stations on the Madeira River in order to have enough food, even using force when the owner resisted. The patrones and boatmen with whom Mathews traveled helped themselves to plantains and pumpkins from the outlying chacos of Exaltación, claiming that this was permitted by custom.[28]

Indian boatmen faced the additional dangers of drowning, disease, and hostile encounters with independent Indians. Travelers often repeated stories of drownings, but seldom ever witnessed one. Franz Keller and Maccheti mention that the Caldeirão do Inferno rapids claimed many lives. A Brazilian government report states that six canoes sank there in 1864, drowning eight persons. Mathews saw one canoe sink near the Teotônio falls, but its four men were saved. Church plucked sixteen Indians from a rock after they had lost their boat in the cachuela of Pederneira, but lost his own boat at another rapid and was nearly wrecked at Caldeirão do Inferno. Other travelers saw boats capsize without any loss of life. Some drownings resulted when Indian crews attempted to shoot the rapids to avoid unloading and reloading cargo. While traveling beneath the high, eroding banks along the meanders of the Mamoré River, boatmen also could be hit by falling trees or buried under a collapsing bank.[29]

Travelers consistently remarked on the swarms of mosquitoes and other insects that tormented them along the Madeira and Mamoré rivers. Maccheti observed how the feet and legs of his crewmen were swollen from mosquito bites. Franz Keller wrote that most of his crewmen did not have mosquito nets, and slept on hides on the ground. Although the connection was far from certain during the early rubber boom, we now know that mosquitoes brought disease, which is mentioned frequently as a danger of the rubber trade. The cachuela zone was especially dangerous because boatmen had to spend a lot of time portaging around the falls, where rock formations provided mosquitoes with tranquil pools as breeding areas. Pinkas reported several different illnesses contracted by workers in the cachuela zone: intermittent fever, continuous fever, remittent fever, bilious fever, beriberi, diarrhea, anemia, dysentery, rheumatism, pneumonia, dyspepsia, and acute gastritis. Intermittent fever (malaria) was by far the most common illness, and yellow fever probably accounts for some of the other types of fever. One of the Kellers' Indian paddlers died of an "intestinal inflammation" at Caldeirão do Inferno. Mathews and several of his crew became ill in the cachuela zone, including a teenaged paddler who died of fever, and an adult paddler who died shortly thereafter of uncertain causes. Fonseca encountered a returning Bolivian crew just above the Beni River's mouth, in which two paddlers were ill and five others had recently died from sickness. Craig suffered from fever, and reported many other cases of fever, other illnesses, and deaths while in the cachuela zone.[30]

Whites and former mission Indians had as strained a relationship in the rubber zone as they did back in the towns of Beni. The merchants who headed Maccheti's expedition had one of their Indian captains whipped with twenty-five lashes for his alleged carelessness in letting a boat capsize in a cachuela, and Mathews beat a crewman for stealing cane liquor. In the 1860s and 1870s, nevertheless, merchants had to be careful not to antagonize their crews, who could take revenge by letting a boat slip over the falls at night, while the owner was asleep on board. Passing through the cachuela zone, the Kellers noted the spot where a Bolivian merchant had been killed by his own paddlers only a few months before. They claimed that Indian crewmen hated whites and had killed their patrones on several occasions. A Bolivian writer, Manuel Macedonio Salinas, agreed that Indian crews frequently deserted during voyages, "not a few times assassinating the merchants." While on isolated portions of the river, a patrón could not enforce any order without the approval of most of the crew. Indian crewmen could disobey their patrones with no consequences, as when the crewmen of Mathews's expedition chose to roast caiman meat in camp despite their patrones' order not to start fires for fear of independent Indians. When more patrones settled on the rivers, there was less chance for Indians to dispose of them or to disobey them, and whites could treat their crews worse. According to American botanist Henry Rusby, Antonio Vaca Díez once disembarked at the Teotônio falls, aimed a gun at the captain, and forced the crew to pass over the falls, just to satisfy himself that it could be done.[31]

## The Lives of Rubber Gatherers

Rubber gathering was less strenuous and less dangerous than river transport. A typical barraca centered on the houses of the patrón and his workers, located along a navigable watercourse. The patrón's house was usually a large, two-story building with outside stairs leading up to a second-story veranda which often extended completely around the building. Such houses were built of various combinations of timber, split palm wood, and palm leaves. Surrounding this headquarters were plantings of food crops: bananas and plantains, manioc, rice, corn, and occasionally sugarcane. Trails led from the patrón's house through the forest to small dwellings, each housing four or more rubber tappers. From these houses several other trails (called *estradas* after the Brazilian usage) looped through the forest, connecting numerous isolated rubber trees.[32]

*Figure* 3.2. An early rubber patrón's house on the Madeira River. From Franz Keller-Leuzinger, *The Amazon and Madeira Rivers* (Philadelphia: J. B. Lippincott, 1875).

The most productive rubber trees grew in areas that were inundated for a few months each year, often from January until March. During such floods, rubber would not be gathered. When and where the ground was dry enough, a rubber tapper walked to each tree on an estrada and made a downward-sloping cut in its bark. Sap oozed from this cut into a piece of bamboo, and thence into a little clay cup, or else directly into a small tin pot hung on the tree. As he made his rounds, the rubber gatherer emptied each full cup or pot into a large gourd, which he carried on his back to his hut where he emptied it into a large turtle shell or similarly sized vessel. Craig found an alternative method of gathering, in which men made cuts into the bark early in the morning, and women gathered the sap around 11:00 a.m. At the hut, the rubber gatherer built a fire of palm nuts, dipped a wooden paddle into the sap, and held it in the smoke, which made the sap quickly coagulate into raw rubber. The fire often had a clay chimney placed over it to concentrate the smoke. The paddle received several coats of sap until the rubber was a couple of inches thick, and then the mass would be cut off the paddle and the process would begin again. Such a mass of rubber was called a *plancha*, a *bolacha*, or a *borracha*. Every Saturday, the tappers would carry their rubber to the barraca

*Figure* 3.3. A later rubber patrón's house: Nicolás Suárez's house at Cachuela Esperanza. Author photograph.

headquarters, and every Sunday, they would return to their outposts with enough provisions for the following week.[33]

Although Mathews claimed that "rubber speculators and merchants descending the rapids will not allow the Indians to take any of the females of their families with them," he also noted that the paddler who died in the cachuela zone left behind an Indian wife on the Madeira River, and that the thirty Cruceño men bound for Baetas had four or five women with them. Craig saw "several hundred Bolivian Indians, with their families,"

*Figure 3.4.* Preparing rubber on the Madeira in the 1860s. From Franz Keller-Leuzinger, *The Amazon and Madeira Rivers* (Philadelphia: J. B. Lippincott, 1875).

at Concepción de Morrinhos, and observed women working in the smaller dependencies of that settlement. At one of these, the wives of rubber gatherers prepared corn chicha the same way as in Bolivia, by chewing the kernels and spitting them into a trough to ferment, while at another, a Bolivian girl performed domestic chores for the eleven men who worked the surrounding estradas. Pinkas wrote that the huts of workers at a barraca were separated by the sexes, with the patrón's house placed between the two groups of huts, and listed 130 women and children among the 520 workers in the barracas of the Río Madre de Dios. Maccheti baptized four or five children that had been born on the Madeira River and returned to Exaltación in Beni with their families.[34]

Not all of the women in the barracas lived stable family lives. Around 1879, Prefect Dr. Benjamín Lens attempted to prohibit the recruitment of Beni Indian women for the barracas, believing that they were used as "bait" to lure Indian men to work in the rubber settlements. Pinkas commented that Beni Indian women would offer themselves to men coming downriver from Bolivia, and that Bolivian patrones kept Beni Indian concubines, sending older women to work the rubber forests when they could replace them with younger arrivals.[35]

The diet of rubber gatherers was similar to that of boatmen, with more locally grown food and less game. In addition to the abovementioned chicha, Craig found manioc, plantains, and chickens (kept for their eggs). Pinkas noted that rubber workers ate manioc flour mixed with water. The Jesuit Gumercindo Gómez de Arteche, who did not visit any barracas, claimed that siringueros only consumed "a little flour and coffee, and at times a little *aguardiente* [cane liquor]." With regard to game, Franz Keller observed that siringueros captured three to four thousand female turtles at the Praia de Tamanduá during the laying season and kept them alive for later consumption. Turtles, he wrote, "take the place of beef in these regions."[36]

The quantity of food varied, and workers in some barracas were very poorly provisioned. In order to increase rubber production, some patrones would not allow their workers sufficient time to grow food, hunt, or fish. Even if workers planted their own crops in a new barraca, they would have had to wait five months for corn or rice to mature, eight months for manioc, and a year for bananas and plantains. In one Beni River barraca visited by Rusby in 1886, both the patrón and his Indian workers were short of food. The patrón had to agree not to touch his workers' cornfield, and they to take nothing from his household stores. Rusby also recorded an incident when Antonio Vaca Díez was running out of food and determined "to dismiss the peons that he could no longer feed, dispersing them through the forest, each to shift for himself as well as he could."[37]

Rubber gatherers did not share the same risk of drowning as boatmen but did have to worry about illness and "bárbaros." In November, when the first floodwaters from Beni reached the Madeira River, fever would sweep through the valley. Franz Keller wrote that when the forest was flooded, a rubber tapper had little to do but "calculate exactly the intervals between his fits of ague." Unlike whites, Indian workers rarely had access to quinine to fight malaria. Their own remedy was a bitter herb called caferana; how successful it was, Keller does not report.[38]

## Construction Work

In the early years of the rubber boom, Beni Indians could take advantage of a journey to Brazil to escape from Bolivian patrones altogether. Brazilians employed Beni Indians on public and private works in Manaus as early as 1864. Whole families were living there by 1865, when they had their portraits taken by the pioneer German photographer Albert Frisch, who had accompanied the Kellers to the Amazon.[39] Franz Keller saw many Beni Indians in Manaus in 1868 and observed:

> They were about the only persons we saw working in the streets, carrying turtles and fuel from the shore to the houses, or lending a hand at new buildings. They here gain about ten times as much as they could in their own country, where they live in great misery; and so there is an endless current of emigration from Bolivia to Brazil, in spite of all the reclamations of the former.[40]

A year later, Maccheti saw many Beni Indians camped out in the plaza of Manaus, and the Brazilian authorities recorded that twenty-eight Bolivians arrived in Manaus, and only sixteen left. The economic incentive must have been enough to draw many former mission Indians to Manaus in the first years of Bolivian rubber exploitation, and a skill such as carpentry learned in the ex-missions could pay as much as four milréis per day in 1870.[41] Their opportunity to settle in Manaus disappeared by 1873 when, with the advent of regular steam navigation on the Madeira, Indian crews were no longer needed below San Antonio.

But in 1873, a new demand for Bolivian labor appeared when the Public Works Company of London began construction work for Colonel Church's Madeira-Mamoré Railway Company. The construction company hired laborers from Bolivian patrones, who left their crews to work in the cachuela zone while they traveled by steamboat to the Amazon. These patrones received money from the company and were to pay their crews afterward, but the company also hired workers directly, paying them twenty milréis per month. The company provided all workers with jerky, beans, rice, coffee, and sugar, and gave them fresh meat twice a week. Forty to fifty Beni Indians were employed unloading and storing

*Figure* 3.5. Mojos in Manaus in 1865. By Albert Frisch, reproduction courtesy Museu de Arte Moderna-Rio de Janeiro, by permission.

construction material brought by steamboat, and in cutting down forest for the railway route. Mathews, the company's engineer, contracted with Ignacio Araúz to provide sixty to seventy more Beni Indian workers to cut trees. The company employed no Brazilians and tried to recruit more workers in Bolivia, but was unsuccessful, and gave up the project later in 1873. When the P. & T. Collins Company of Philadelphia took up construction work on the railway in 1878, they hired 300 Bolivians along with 500 Americans and Italians. Later that year, 120 Italians left, but a new workforce arrived with 500 northeastern Brazilians fleeing the Great Drought in Ceará. The Collins Company abandoned the project in 1879, and Church lost his concession from Brazil in 1881. Bolivian workers, so essential for the 1873 attempt, were outnumbered by Brazilians by 1878, a sign of declining opportunities for Bolivians on the Madeira. When the American Percival Farquhar's new Madeira-Mamoré Railway Company finally succeeded in laying rails around the cachuelas in 1907–12, most of its 22,000 workers came from overseas—the West Indies, India, and southern Europe.[42]

Disease haunted the cachuela zone and was responsible for the failure of the first two construction attempts. One Brazilian account notes how unvaccinated Beni Indians suffered from smallpox at San Antonio on the Madeira in 1873. Beni Indians helped unload construction material for the Madeira-Mamoré Railway from the steamboat *Amazonas* and, after coming into contact with an infected person on board, four became ill and three died. The patrón Fidel Hurtado tried to return to Bolivia, but most of his crew became infected, and he had to turn back after seven crewmen died. In the end, the railway company vaccinated twenty Beni Indians at San Antonio to stem the spread of smallpox among them. The 1878–79 attempt was even more deadly. According to historian John Hemming, 200 Bolivians died along with 400 Cearenses and 141 Americans, mostly from malaria.[43]

## Culture in the Rubber Zone

Both boatmen and rubber gatherers were able to maintain a fair amount of their hybrid Christian-indigenous culture, especially before 1880. Paddlers dressed in wide-brimmed straw hats and long bast (tree bark) shirts for the workday and changed into cotton camijetas when they went to sleep; both types of cloth were used in the former mission towns. Mojos continued to wear their distinctive handmade shirts and straw hats on the Madeira River, in the streets of Manaus, and on the lower Amazon. Franz Keller's paddlers each had two or three handmade rosaries, and one Trinitario paddler even carried an old prayer book written in Mojo. Even before the rubber boom, Orbigny noted that Indians on a river voyage would pray and sing canticles after dinner. Similarly, Maccheti's crew began their day with religious songs and prayers, while Mathews's ended theirs with songs whose refrain was "deliver us always from every ill."[44]

Concepción de Morrinhos had a large wooden cross like those in the plazas of the former mission towns of Beni, facing inland near the riverbank. "The cross was about eighteen feet high, neatly painted black, lettered and tipped in white and from the arms hung two pieces of lace about eight feet long and nine inches wide," wrote Craig. The whole Bolivian Indian population prayed and sang before this cross at sunset. They also had a band that played violins, drums, flutes, and triangles. The Indians maintained some of their festivities as well, including dancing, singing, and "a sham bull fight . . . the bull being an Indian covered with the head and hide of an ox." Rather than an attempt to re-create the bullfights of Beni feast days, this was likely to have been a traditional torito dance, in which one dancer always dressed as a bull. There is some evidence that Beni Indians in the rubber zone still believed that they could communicate with their ancestors. The wife and relatives of Mathews's crewman Mariano suspected that he would die and asked Mariano to carry messages from

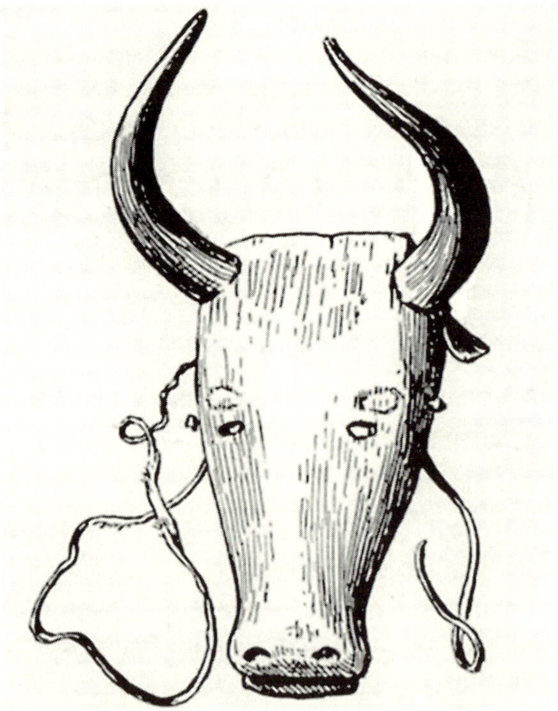

*Figure 3.6.* Baure Torito dance mask. From Erland Nordenskiöld, *Indianer und Weisse in Nordostbolivien* (Stuttgart: Strecker und Schröder, 1923).

them to their ancestors. Mathews claimed to have witnessed such requests several times.[45]

## Relations with "Bárbaros" in the Rubber Zone

Relations with the independent Indians could be either hostile or friendly. On the Madeira River, the Muras lived far downstream and were no longer considered dangerous by Franz Keller's time. Upstream were the Araras, whose presence discouraged explorers from following right-bank tributaries of the Madeira. Further upstream, the Parintintins prevented settlement by outsiders on a short stretch of the river. A Brazilian report recounts how Parintintins killed two women with arrows at a place called Frexal in 1868; one was Bolivian. Franz Keller wrote that Parintintins had burned a house and killed a family near Crato, and that they attacked both Englishmen and Mojos working on the Madeira-Mamoré Railway, killing two Mojos. Craig and Mathews both heard of the same Parintintin attack near Crato. Brown and Lidstone commented on the hostility of the Araras and Parintintins, and Fonseca reported that Parintintins had killed the crew of a boat ascending the Madeira in 1876.[46]

Travelers who wrote about journeys through the cachuela zone always mentioned the danger of attack by local indigenous peoples, who included Caripunas, Pacahuaras, Araras, Acanga Pirangas (Uru-Eu-Wau-Wau), and Parintintins. Despite the reputation of some of these groups for hostility, only Craig reported actual attacks that occurred while he was in the area. The Caripunas invited Keller and his crew to their village near the cachuela of Caldeirão do Inferno, and traded sweet manioc, corn, bows, arrows, and feather ornaments, for knives, scissors, fishhooks, white glass beads, and red cotton handkerchiefs. Maccheti's crew traded with independent Indians ("Araras," but perhaps really Caripunas) near the cachuela of Araras, who wanted shirts and American hatchets. They actually received horn spoons and combs, old shirts, bark for new shirts, and straw hats, all prepared by the crewmen, as well as aguardiente and salt, in exchange for their corn, plantains, and manioc. By 1878 there were a few Caripunas living alongside Bolivians. Some of the four Caripunas who lived with Ignacio Araúz and his workers at Três Irmãos had adopted western clothing and Spanish names.[47]

Travelers reported very different relations between the former mission Indians and the "bárbaros." Franz Keller described how his crewmen regarded independent Indians with "a curious mixture of fear and contemptuous disgust," and how his Mojo steersman disliked non-Christian

*Figure* 3.7. Beni Indians meeting Caripunas on the Madeira in the 1860s. From Franz Keller-Leuzinger, *The Amazon and Madeira Rivers* (Philadelphia: J. B. Lippincott, 1875).

Indians. Pinkas also related that Beni Indians were afraid of independent Indians and had no desire to encounter them.[48] This, at least, is what the former mission Indians wanted whites to believe. In telling Keller and Pinkas of their fear and disgust, the Beni Indians set themselves apart from the "bárbaros" on purpose, because they knew that whites respected "civilized" people more than "savages."

Other travelers saw how fluid the boundary between civilized and savage really was. Maccheti observed very amicable relations between ex-mission and independent Indians. The independent Indians he visited could make the sign of the cross and adopted Christian names. They knew some Spanish and Portuguese, but actually had a much greater knowledge of the Cayubaba language, "because when these [Cayubaba] Indians come here as crewmen, they desert and live for whole years among these bárbaros, until they are presented with a favorable occasion to return to their town." On Maccheti's own expedition, three Cayubaba crewmembers stole a sack of flour and fled to the independent Indians. The Cayubabas, at least, thus appear to have established a special friendship with one group of independent Indians that enabled them to escape the authority of whites whenever they pleased. Several years later, Mathews encountered a group of Pacahuaras near the rapid of Três Irmãos whose chief, Mariano, was a runaway Cayubaba from Exaltación in Beni. They spoke a few words of Spanish and Portuguese and were also fond of taking Christian names from the travelers who passed by.[49]

Travelers did report several attacks by independent Indians in the cachuela zone, though as with drownings, these were mostly secondhand accounts. At the falls of Girão, an old mulatto in Franz Keller's crew told him that he and his companions had hired Caripunas to help portage in 1860. In a dispute over payment, the Caripunas killed one companion and wounded two others with arrows, and several Caripunas were wounded by gunfire in return. Many reported attacks happened on the stretch of river between the rapids of Morrinhos and Caldeirão do Inferno, where Keller and Mathews observed that the Caripunas had settlements. Maccheti recounted how nine months before his voyage, a Bolivian named Montaño and several companions were left behind there by their crew when Montaño refused their demands for more payment. Montaño and most of the others were then killed by independent Indians. Craig heard of a Caripuna attack there in April 1875 which left five of seven Bolivian traders dead. During his own stay around the rapids, two Bolivian Indians and one American were killed, and three Bolivian Indians and two Americans wounded, by unknown "savages," either Parintintins or Acanga Pirangas.[50]

The Moré or Iténez Indians, who lived near the confluence of the Mamoré and Guaporé (Iténez) rivers, were regarded as hostile by most travelers. Franz Keller reported that Indians (apparently Moré) had killed Brazilian soldiers near Forte Príncipe da Beira on the Guaporé, killed and wounded several Cayubabas who had descended the Mamoré to gather cacao, and kidnapped a Bolivian steersman who had gone ashore for gulls' eggs. Maccheti heard an account of a probable Moré attack on a Brazilian boat near the Iténez River mouth in 1869, which resulted in the death of a Brazilian consul and several crewmen, and which Franz Keller and Mathews later repeated. Siringueros who settled along the Iténez River were also in danger. In 1887, according to a member of the municipal council of Magdalena, "more than a hundred savages . . . shot more than 300 arrows at the workers, having caused the death of a woman and grave wounds to others among them."[51] Independent Indians attacked individuals or small, relatively defenseless groups of strangers, and traded with large, well-armed expeditions of many boats. Both types of behavior can be explained by their desire to obtain manufactured goods; whether they used force or traded peacefully depended on the strength or weakness of the foreigners who passed through their territories.

## Recruitment of Labor

Many accounts note that boatmen were recruited through the carayana and indigenous officials of the former mission towns of Beni. Franz Keller wrote that the corregidor of Exaltación and the prefect of Beni obtained paddlers for river transport. Both worked through the cabildo leaders below them, as the "chieftains" (caciques) of the towns would hire paddlers and accept the crews' payment in advance. The Bolivian politician Ladislao Cabrera was only able to obtain a boat and paddlers through the prefect and the corregidor of Exaltación. Salinas wrote in 1871 that paddlers from Beni could only be obtained with difficulty, and only through the authorities. José Gregorio Acuruza, a Cayubaba cacique who at times was corregidor of Exaltación, assigned many of his people as paddlers to merchants and rubber patrones. These beneficiaries gave him many gifts with which he decorated his house, including many "vistas of country-houses and fields of Europe placed in great gilded frames," which Acuruza claimed were views of his own properties in Beni.[52]

The wages of Beni Indian boatmen actually climbed as labor grew scarcer during the early rubber boom. Mathews wrote that they got 30 to 40 percent more in 1874 than they had a few years before. For a trip from Trinidad to Coni on the route to Cochabamba and back, they used to get six to seven pesos, but received ten in 1874. In 1889, Cabrera paid crewmen making the same voyage twenty-eight pesos each. The official wage in 1888 and 1889 for paddlers on the mail routes from Trinidad to Santa Cruz and Cochabamba was only eight pesos, but even this was an improvement over the unpaid service ordered by the national government in 1856.[53]

While in their home towns, Beni Indians could refuse to start a river voyage just before or during a feast day, and would make their patrones wait a few days after the fiesta in order to have time to recover from heavy drinking. When away from home, Beni Indian crews could hire themselves out to the highest bidder, or to the merchant or traveler whose itinerary would take them home soonest. In the rubber zone, Indian kinship networks played a role in the recruitment of boatmen. Mathews noted that "where practicable, the plan is to find a good captain and let him select his men from amongst his own 'parientes' or relations."[54]

As rubber gathering was easier than paddling, some Beni Indians abandoned river transport in the rubber-collecting region to work in the barracas. Just before Mathews's expedition entered the cachuela zone, a Bolivian boy hid in the woods in order to be left behind. Mathews's interpretation was that "there are many of these Indians that are so lazy that they would rather remain on the Madeira River than undergo the hard work of the ascent of the rapids on the way to their own country." Fonseca encountered a Bolivian boat that had been stuck near Caldeirão do Inferno for six months after several of its paddlers deserted. Salinas wrote in 1871 that six hundred Beni Indian paddlers had chosen to abandon their families in Bolivia and remain in Brazilian territory. Rubber patrones also made arrangements with merchants to acquire paddlers as laborers in the rubber zone. One of the cinchona merchants with whom Maccheti traveled left two of his crewmen to work for Ignacio Araúz at Exaltación on the Madeira.[55]

Rubber patrones who owned ranches in Beni, like Rómulo Suárez in Trinidad and Pastor Oyola in Concepción de Baures, could have sent peons whom they already controlled to work in the rubber zone. When Craig visited Oyola at Concepción de Morrinhos, he noted that "Oyola's wife was absent on a long and tedious journey to Bolivia in quest of peons and to visit friends." Whether she recruited labor from the Oyola properties or through local Indian cabildos is unspecified, but significantly, Oyola's wife was a Baure of the elite Ojopi family.[56] This marriage alliance between Pastor Oyola and the Ojopis, with their ranches and connections to the cabildo, facilitated the transfer of labor from Concepción de Baures to its namesake, Concepción de Morrinhos.

## Enganche and Debt

Contemporary and later Bolivian writers often use the term *enganche* to explain how labor was recruited for the rubber boom. Modern scholars usually use the term *enganche* to mean the advance of money (or perhaps alcohol) by patrones to workers as a way to trap workers into a cycle of debt that they could never hope to work off. But contemporary writers such as the Kellers and Salinas used "enganche" to refer to the recruitment of workers through the corregidores and cabildos of former mission towns. Thus there is little reason to doubt that this was the principal means by which Beni Indians were sent to work in the rubber zone either as boatmen or rubber gatherers. The use of enganche (in its scholarly sense) to recruit workers applies more to Cruceño mestizos than to Beni Indians. In the "houses of enganche" of Santa Cruz, labor recruiters lured potential workers in with music and alcohol, and one house bore the sardonic sign "Beni Street, where you go and don't come back."[57]

Franz Keller wrote that rubber workers were held in a "cleverly designed bondage" because of debt. Their patrones hardly ever paid them cash wages, "but rather in goods and provisions charged at thrice their value," while crediting the workers with less than half the export value of the rubber they produced. Mathews also claimed that rubber patrones held their Indian workers in a kind of slavery, keeping them bound to their barracas "by means of debt and drink." On Sundays, the workers would deliver a week's worth of rubber to the patrón, who would then get the Indians drunk on cane liquor. While drunk, an Indian would be "induced to buy trinkets, calicoes, ribbons, and other articles that he could very well do without" from the patrón. The Indians were overcharged for these goods and credited very little for their rubber, and so being constantly indebted to the patrón, they could never leave. Pinkas also described rubber workers at Bolivian barracas on the Madeira as debt peons and called their bondage "a new slavery." He said the workers were recruited in Beni, and that the men were to receive fifteen milréis (about 1.5 pounds) and the women ten milréis (about one pound) per month. However, the patrones charged their workers for transport to the rubber zone, for pants and shirts for the men, for tipoys for the women, and for other "trifles," so that they would immediately be in debt for one hundred milréis (about ten pounds). The workers accumulated more debts, sometimes of two thousand to three thousand milréis (about two hundred to three hundred pounds), and therefore could never leave their patrón.[58]

Other observers of the rubber boom elsewhere in Amazonia shared this impression that rubber workers were bound into perpetual servitude by debt. The works of Barbara Weinstein and of Bradford Barham and Oliver Coomes, however, make it clear that the debtor relationship of rubber workers with patrones cannot be equated with slavery. In most cases (the Putumayo being an exception), patrones could not really oppress their workers, as labor was scarce and workers could easily run away, either to another patrón or to the cities of Manaus or Belém, where even unskilled laborers could receive high wages. Nor could patrones keep an eye on the activities of their highly dispersed workforce. At least below the cachuelas, workers could have exchanged some of their rubber with itinerant river traders for more goods than the patrón offered. Both boatmen and rubber gatherers are said to have occasionally accepted advances from several patrones and then disappeared.[59]

The economy of most of Amazonia in this period was so cash-poor that an exchange of provisions, clothes, and other goods for raw rubber made more sense than paying a wage. Weinstein suggests that a principal reason patrones kept their workers in debt was to avoid having to pay them in cash. It is true that workers were charged greatly inflated prices for goods, and that the credit they received was much less than their rubber eventually sold for in Belém. But though it may have existed on the books, a debt meant little in reality given the scarcity of cash. Even patrones would accept goods from merchants instead of cash. In return for the rubber that patrones shipped downstream, their agents in Belém were always ready to send them luxury goods, such as the silks, silk umbrellas, and Panama hats that Pastor Oyola had in his house. A chain of credit thus extended from trading houses in Belém through river traders to patrones on the Madeira, and from patrones to their workers. The firm of Suárez Hermanos avoided a debt relationship with traders by having its own agencies handle every step of the trade from Cachuela Esperanza to London, but such a case of vertical integration was rare.[60]

There is little evidence that debt or force kept Beni Indian workers at the Madeira barracas. However, life in the barracas provided an opportunity to continue living as a community, with lighter demands on their labor than in Beni, and no obligation to pay taxes and tribute. In Pastor Oyola's settlements, with a large cross for prayer, torito dances, and the making and consumption of chicha, Beni Indians could maintain some of their cultural life. Furthermore, several barracas were named after former mission towns or their inhabitants, suggesting that the population of each settlement was predominantly of one ethnic group.[61] Some Beni Indians may have preferred community life in a barraca to community life in a former mission town for both cultural and economic reasons. Once in the rubber zone, the only other options besides trying to return home would be life in a Brazilian city, or life among the "bárbaros."

It should be noted that the scenario of relatively good working conditions advanced by Weinstein, Barham, and Coomes is best applied to conditions on the Madeira River below the cachuelas, before 1880. By 1880, the disadvantages of participation in the rubber boom came to outweigh the benefits for Beni Indians. Regular steamboat service, established on the Madeira by 1873, lessened their chance of reaching Manaus during their regular voyages. Patrones started demanding more labor to exploit the richest rubber forests, which were upstream from Brazil and offered less opportunity to escape. As more patrones settled along the rivers, Beni Indians had less room to escape to independent groups, or to disobey or dispose of their patrones.

## The Rubber Boom and the National Government

The Bolivian national government had always expected its officials in Beni to glean their salaries from local resources—first temporalidades production, and then state-owned cacao groves or wild herds of cattle. It is little surprise, then, that when newly rich merchants or rubber patrones offered money for labor, the local authorities would act as recruiting agents. The national government, alarmed at the decline in tribute from Beni, ordered that a bond of ten pesos be paid for each Beni Indian taken to the Madeira River in its *Reglamento* of May 5, 1871. The same ruling exempted the Trinitario Mojos from recruitment as rubber workers, reserving them instead for road work around Trinidad. This reglamento was poorly enforced, as the Trinitarios "complained bitterly of the great emigration to the rubber-grounds of Brazil" to Mathews in 1874.[62]

The national government had so little control over Beni that it soon had to make an unprecedented concession to the local elite during the War of the Pacific, which began when Chile attacked Bolivia and Peru and seized the nitrate-rich Atacama Desert in 1879. On May 26, 1880, the Bolivians fought and lost their last battle of the war, and then left their seacoast in Chilean hands. President Narciso Campero realized that Beni offered access to foreign trade, including arms, now that the Pacific was inaccessible. In a supreme decree of July 20, 1880, the government agreed to consult an "assembly of notable citizens" of Trinidad when choosing the prefect of Beni. Bolivia was officially still at war, and this decision was born out of national crisis and was uncharacteristic of Bolivia's centralist tradition of appointed prefects. It also essentially turned the prefecture over to the rubber interests. Although it is unclear how long this decree remained in effect, Beni prefects Dr. Mamerto Oyola (1881–86) and Daniel Suárez (1886–87) came from the Cruceño elite. In another demonstration of rubber men's rise to political power, the powerful rubber patrón Antonio Vaca Díez was elected senator from Beni in 1884.[63] Such political power streamlined exploitation and greatly lessened the chances for Indians to play off the national government against the local white elite.

The rubber boom did create a new source of income for the departmental and national governments. A law of 1883 created a five-boliviano tax on each estrada, which went to the departmental treasury until the tax was canceled in 1887. By another law of 1883, the national government imposed an eighty-centavo duty on each arroba of rubber exported from Bolivia, which it began collecting after establishing a customs post at Villa Bella at the juncture of the Beni and Mamoré rivers the following year. This anticipated income allowed the national government to abolish Indian tribute in Beni in 1883 as well. With the change in income from tribute to rubber export duties, the state's fiscal interest shifted from protection of Indian tributaries to promotion of the rubber industry.[64]

## Punishments in the 1880s

Arbitrary physical punishments against rubber gatherers may not have been common on the Madeira before 1880 because workers could easily flee their barracas. After Bolivian rubber patrones shifted their focus upstream in 1880, conditions worsened and options decreased for Beni Indian workers. Only then is there ample evidence that poor working conditions in some barracas drove workers to run away. On the Beni River in 1881, twelve workers fled from their patrón Fidel Endara after he drunkenly shot off his revolver, hitting one worker in the arm. In the same year, fourteen of Antonio Vaca Díez's Cruceño workers fled for an unknown reason. Pinkas noted the case of an old woman who attempted to escape from her patrón, who pursued her and promised to punish her with fifty lashes. While traveling in the cachuela zone in 1889, Cabrera met a boatman who had previously been a rubber gatherer on the Beni River. After two years in a barraca, this man and two companions fled because of hunger, as they only received one meager meal per day. Their patrón chased them down with a crew of black Brazilians who caught the two other workers and shackled them. The man Cabrera met was able to escape to the forest after menacing the Brazilians with a knife.[65]

Although barracas were located in remote places and their workers were often unsupervised, escape was not as easy as one might suspect. Simply heading into the forest would require that the escaping peon either have friends among independent Indians or a knowledge of which wild plants were edible. Otherwise, escape meant traveling by river, and patrones could easily chase those who attempted this route, as evidenced in the cases cited above. Furthermore, patrones punished those who attempted to escape with whippings. In a barraca on the Beni River, Rusby witnessed the treatment of a man and a woman who had tried to escape. Each was stripped, laid face down on the ground, and had their limbs tied to stakes. "The whip at first produced only great welts, but later cut into the skin and drew blood freely," commented Rusby.[66]

It should be emphasized that all of these recorded attempts to escape and subsequent punishments occurred on the upper rivers in the 1880s. Indian options decreased when the rivers became more thickly settled, when escape to Brazilian cities became more difficult, and when rubber patrones rose to political power, so that whites could have treated their workers worse than before. Life on the Beni, Madre de Dios, Orthon, Tahuamanu, and Manuripi rivers in the 1880s was certainly harsher than life on the Madeira in the 1860s.

## The Cultural Environment in the Former Mission Towns

Life in the former mission towns during the early rubber boom was marked by much continuity with the past. Franz Keller observed many of the same rituals in 1868 that Orbigny and Gibbon had seen earlier in the century. The former mission Indians still attended church regularly and sung many old hymns, and musicians played in church and at festivals. Their instruments were the same as those of the Jesuit period: violins, cellos, flutes, harps, organs, and *bajones* (huge panpipes made from palm leaves). Macheteros danced from cross to cross in the plaza, and before the church's altars. Penitents occasionally appeared in processions, including a crawling man tied by both arms and one leg to a cross, and men and women walking on their knees. Festivals still ended in drinking bouts of corn chicha or cane liquor, and drunken Indians climbed into a palisaded bullring set up in a corner of the plaza to fight a wild bull with knives. The prefect and other officials viewed the bullfight from the portico of the former colegio, but the prefect now threw handfuls of copper coins into the crowd instead of the corn and manioc bread witnessed by Gibbon. And Indian women now went to church in dresses made from "gaudy, large flowered cottons of Europe" rather than their traditional white tipoys of locally made cotton cloth. Both coins and imported cloth reflected the growing commercialization of the Beni economy. In another change, the number of men performing the machetero dance decreased. Certain older Bolivians told Franz Keller that dances used to have many dozens of macheteros rather than the one dozen he observed, and that "they probably would have ceased altogether if the chieftains [caciques] did not exert the full weight of their authority in behalf of keeping them up, even forcing the young men, in case of need, to take part in them."[67] In this case, the cabildo acted as defender of the "traditional" culture (whose tradition was a construct of the mission period).

Mathews described Trinidad in 1874 as a small town of brick or adobe-walled houses with tiled roofs, with only one street left dry during floods. Schools continued to exist. The few important merchants and shopkeepers were carayanas who made their living by exporting cacao to Cochabamba and Santa Cruz, and importing flour and potatoes from Cochabamba, and British textiles from Corumbá via Santa Cruz. There were some mestizos (called *cholos*) in the town, and though most of the population was still Mojo, mestizaje between whites and former mission Indians was increasing. Mathews wrote that the Trinitario Mojos "are becoming so mixed with the carayanas, that complexions of all shades, from almost white to dusky red copper, are found amongst them." The Indians still prepared chicha for their fiestas, when they drank a great quantity of it, and still respected the authority of their caciques, who Mathews said were "generally in the pay of the prefect or corregidor, as the case may be." Mathews saw the fiesta of the Virgen del Carmen in Trinidad, with its accompanying drinking, and noticed macheteros dancing in the church of Exaltación. Franz Keller and Mathews observed that the Trinitarios and other Beni Indians manufactured cotton macanas, bark shirts, dyed-rush mats, and straw hats. The macanas remained in great demand in the highlands, but the Indians were required to sell them at low prices, while they had to pay six times the fair value of imported European textiles. In 1868, the rubber patrón Ignacio Araúz noted that Beni Indians produced cotton hammocks, towels, and clothes for the national market, and ponchos that were exported to Argentina. Franz Keller, however, claimed that both the quantity and quality of macanas were declining by the same year.[68] Beni Indian artisan production had been in decline since the 1830s and would reach even lower levels by the late 1880s.

Men of the leading native families were still important enough to play a role in carayana politics. In the years 1870 and 1871, the downfall of President Mariano Melgarejo triggered a series of coups and countercoups in Beni. When Dr. Benjamín Lens took the prefecture and asked the townspeople to confirm him in a vote, the Mojos Belisario Nosa and Bartolomé Teco joined the dozens of carayanas who signed. Teco was certainly a member of the Trinitario elite, having served as sacristan in 1861, while Nosa was probably also elite, since at least three other Nosas had held high offices including corregidor, cacique, and músico.[69] Although carayanas had stopped calling elite Trinitarios citizen by the 1870s, they still regarded some of them as near-equals.

Bolivia's preeminent historian, the Santa Cruz–born Gabriel René Moreno, offered his interpretation of Trinidad society in the 1880s:

> Today Trinidad has tile-roofed houses and an upper, vecino class of whites and near-whites. Its social structure tends to be similar to that of the other towns of Bolivia: Indians, Indian-white mestizos, [and] Spanish creoles comprise the sum of its inhabitants. Among the towns of Mojos, it is the one where the fluid of nationality has been able to transfuse its most proven and peculiar outpourings. The Bolivianism of our times is already filtering into that social container in blood and spirit. The natives have been intermarrying with the collas [highland Bolivians] and Cruceños more than a little, since thirty years ago. Pure Mojo blood is

> diminishing in the veins of those born there. The camijeta and tipoy, the mission Indian costume par excellance, are disappearing in Trinidad. There, whoever feels the spirit that even four drops of Caucasian blood give, adopts the jacket and pants and the skirt and shawl of the common people of Santa Cruz. So, the individual is in the position of receiving and propagating the Bolivian fluid itself through heredity, even when his body still reeks of camba [lowland Indian peasant].[70]

In Moreno's opinion, then, the Trinitarios were becoming more Bolivian and less Mojo, as the "fluid of nationality" entered their veins through intermarriage, as a property of "blood," which could then be passed on to future generations. Mestizaje was thus the key to creating a common nationality, which was expressed by an immediate change in clothing. Yet it is unclear whether this was a good thing to Moreno. A member of the Cruceño elite who spent much of his life exiled in Chile, his language reflects a conflicting mixture of old white vecino attitudes and newer European racial ideas he had encountered in Santiago de Chile. The European theorist in him viewed any admixture of "Caucasian blood" and any adoption of European-style dress as improvement, while the vecino in him disdained what he saw as smelly Indian peasants trying to pass as white by dressing as Cruceño commoners. What is obvious is that Trinitarios were intermarrying with carayanas and changing their style of dress through their own initiative, and that they were joining the national society more than the natives of other Beni towns were.

In 1887, Gumercindo Gómez de Arteche visited the towns of the Mojos and Canichanas. He noted that the Indians would get up early in order to pray and sing the Ave Maris Stella at dawn on their knees in front of the plaza cross or the church. They would perform similar rituals at noon and 8:00 p.m., with the boys praying at nightfall as Gibbon had observed in 1852. Everyone attended church on Sunday and sang the litany and the Angelus. Arteche also provides a more fine-grained commentary than Moreno on changing Trinitario dress. Men and women went to church wearing their cotton tunics and wooden rosaries, and always bathed and changed their tunics beforehand. When going to work, some men had begun to wear pants and a shirt in place of the traditional cotton tunic or bark shirt. In addition, Arteche claimed that the Indians' religious customs were beginning to fall into decadence because of a shortage of priests and because of the "ridicules and tauntings that they suffer from the carayanas." At one point, the Jesuit says, the carayanas even prohibited the machetero dance "as contrary to the culture of these times."[71]

It is hard to estimate the adequacy or inadequacy of Indian diets in the former mission towns of Beni during the early rubber boom. Mathews claimed that the standard monthly wage in Trinidad was five pesos, or sixteen shillings, per month in 1874. This was less than half of what the lowest paid workers in Britain made around the same time. Still, a Mojo could have bought either about 83 pounds of manioc flour, 167 pounds of unhusked rice, 125 pounds of fresh meat, forty-two pounds of jerky, fifteen bottles of aguardiente, or 667 cigars per month with that amount of money in that year.[72] The actual amounts of food available for individuals' subsistence are more difficult to calculate, as they had to pay taxes and tribute, they may have been in a debt relationship with a patrón and relied on advances of food, or may have had access to food from their own fields.

By 1887, Arteche could note "the death of the parcialidades, that is, of the different arts that the different groups of families exercised with skill." Among the Mojos and Canichanas, he found only one Indian blacksmith in San Pedro, and "an occasional loom in the different towns" which could still produce a few fine cotton textiles. An official report from Magdalena in the same year observed that the importation of "foreign macanas" had helped cause a decline in native textile manufacturing. In Arteche's account, the only "industries" which still persisted with much vigor were those related to the church, which carried the most prestige as they were listed before the manual arts in the padrón of 1854. According to Arteche, the parcialidad of Músicos was still performing its function in San Ignacio in 1887, but had not survived as well in the other Mojo towns or in San Pedro. In all the Mojo and Canichana towns, however, the parcialidad of Sacristanes continued to fulfill its duties and was monopolized by members of the most important Indian families. Sacristanes assisted in religious services with the aid of *abadesas* (abbesses), who "had to be women venerable for their age and for their virtues."[73]

In Arteche's view, "the death of the parcialidades" was a direct result of the rubber boom. First, the labor needs of the rubber boom required much less varied skills than a mission community, so that "nowadays they only practice paddling and throwing lassoes at bulls." Second, the parcialidades were based on families, and the rubber boom destroyed families. Beni Indian men married and had children at a young age, and were often husbands and fathers by the time they were sent to the Madeira. Their families might receive no further news of them, and the wives, not knowing whether their husbands were alive or dead, would be constrained from remarrying. Although there were Beni Indian women in the barracas, they

*Figure 3.8.* Old Baure woman. From Erland Nordenskiöld, *Indianer und Weisse in Nordostbolivien* (Stuttgart: Strecker und Schröder, 1923).

were almost always far fewer in number than men. Arteche noted that the women and children who remained behind in Beni tended to fall under the control of patrones.[74]

Arteche's report raises several questions. The decline in artisan production that he attributed to the rubber boom had actually begun as early as the 1830s. It is noteworthy that the newer manzana names are nowhere mentioned; these were clearly less relevant to the Indians' lives than the older parcialidad divisions. But Arteche confused the parcialidades of Músicos and Sacristanes with those who actually served as musicians and sacristans, when as we have seen, they were not the same as early as 1861. It is perhaps better to conceive of musicians, sacristans, and abadesas as small groups of people from various parcialidades who served the church. Still, Arteche makes it clear that the office of sacristan was only open to men of the leading Indian families, and that they did not have to go to work as boatmen, siringueros, or cowboys.

Such privileges recall the difference between the familia and pueblo orders of Jesuit and later times. However, the replacement of the parcialidades, with their differentiated ethnic and occupational names, by the manzanas, with their uniformly liberal-patriotic names, makes it difficult to trace the continued existence of pueblo and familia orders in Beni Indian society. It is unlikely that the familia survived intact, since the artisans, formerly members of the elite class, essentially became commoners when native industry declined. In place of the former division between familia and pueblo, a new social order emerged in which the principal division was between cabildo members (especially caciques) and their families on one hand, and common people on the other.

There is evidence that the cabildos of the former mission towns were involved in the exploitation of their own people. The best-documented indigenous leader is José Gregorio Acuruza, the cacique and occasional corregidor of Exaltación, who was visited by Mathews in 1874 and by Cabrera in 1889. According to traditions recorded by local historian Emma Banzer Toro de Añez, Acuruza was the mestizo son of a Cayubaba mother and a carayana father. Born in 1815 or 1816, he took his mother's surname, an indication that it was still more prestigious to identify with an elite Cayubaba family than a carayana one. He served as cacique as early as 1854 and continued in that post in 1861. Mathews found Acuruza living with many Cayubabas at an hacienda five miles down the Mamoré from Exaltación. These were mostly women and children; the women outnumbered men by five or six to one, presumably because the men were off doing rubber-related work. Acuruza lived in a very large house similar to those of the rubber patrones on the Madeira. It was large enough to provide sleeping quarters for the women and children in "certain great cupboard-like constructions," as well as an oxen-powered sugar mill and space for butchering animals, cooking, and preparing chicha, all under one roof.[75]

Cabrera wrote that Acuruza "governed the town despotically, not so much as corregidor but rather as the astute native that knows the resorts necessary in order to exploit the ignorance of those like him." He had little respect for the prefect's authority, saying that "the Prefect governed in Trinidad, and he in his town." Acuruza had removed all the indigenous families to his estancia El Pilar outside of town, where the women farmed and manufactured cloth for him, and the men farmed, tended cattle, and above all traveled the rivers for him. Boys became paddlers by age eight, and the oldest doing this work was twelve. Acuruza gave women and children food and clothing as payment, paid men a trifling amount of money, and made use of the whip on all of them. All of his workers called him Tata (father) Acuruza and removed their hats when passing his house, even if the doors were closed. In essence, Acuruza continued the compulsory labor of the temporalidades system at a local scale. "With these conditions he has made and makes a great fortune," observed Cabrera.[76]

Many caciques took advantage of the opening of land to private ownership in 1842–56 to acquire large private ranches of their own, with Indians of the pueblo order becoming their peons. Acuruza's El Pilar was one example of this, but caciques in Baures, Magdalena, and other towns also acquired large amounts of land and labor for themselves. These caciques from the Ojopi, Omireji, Oniaba, Avaroma, Cholima, and Guacama families furthermore began or continued "whitening their lineages through marriages with whites and mestizas," sometimes linking themselves directly to rubber patrones. As already noted, Pastor Oyola married a Baure woman of the Ojopi family. Nevertheless, we must remember that marriages between Indian elites and carayanas predated the rubber boom. Acuruza was a mestizo, as were the children of the Baure Hipólito Ojopi and the Cruceña María Manuela Vaca.[77]

## Population Decline in the Former Mission Towns

Two types of sources provide information on the population of the former mission towns. Bolivian government records, mainly dealing with Indian tribute, are the principal source of quantitative data. Accounts of Bolivian, Brazilian, European, and North Americans travelers include valuable observations about Beni towns. Evidence from both types indicates that the Indian population of Beni declined in the late nineteenth century. The Kellers blamed the growing Madeira River trade for the depopulation of Beni, saying that "part of the Indians remain in Brazilian territory and another part pay for the fatigues and privations [of river voyages] with their lives." By counting abandoned houses, they estimated that Exaltación had lost about twelve hundred people out of three thousand. Mathews wrote that both Exaltación and Trinidad appeared depopulated, and that "the reason for the decline of the Indian population is to be found, without doubt, in the baneful effects to Bolivia of the rubber trade of the Madeira and Purus rivers."[78] The Jesuit Arteche, visiting the Canichana town of San Pedro in 1887, observed:

> The town is nowadays reduced to the houses on the plaza, and where before 800 to 1,000 marriages were counted, one only finds 50 to 60,

> and these scattered throughout the countryside; children more than 8 years old are scarcely ever encountered, and the reason is because the carayanas take their fathers willingly or by force, giving them 20, 30, or 40 pesos at most. The cause of the almost complete destruction of this town, as well as the notable diminution of the others, is the great number of people who have died on the Madeira collecting rubber, taken there by the carayanas, almost always by force.[79]

Bolivian government statistics largely support the impressions of travelers. Changes in the recorded tributary populations of the cantones of the Department of Beni are summed up in table 3.1. From these statistics, we can ascertain that the number of former mission Indian tributaries living in their home cantones did in fact decline during the early years of the rubber boom. While they left their homes, this does not necessarily mean that most of them died, as some authors have claimed. Indeed, many of those tributaries who left continued to be enumerated in the lists of their home cantones as ausentes, or absent tributaries. By 1874, the number of ausentes in the four Mojo cantones had risen dramatically when compared with previous counts (see table 3.2).

**Table 3.1** Tributary originario populations in the Department of Beni, 1849–74

| *Cantón* | *1849* | *1854* | *1867* | *1874* | *% Change 1849–67* | *% Change 1867–74* |
|---|---|---|---|---|---|---|
| Carmen | 213 | 205 | 189 | 92 | −11 | −51 |
| Baures | 734 | 680 | 385 | 164 | −48 | −57 |
| Huacaraje | 217 | 148 | 155 | 96 | −29 | −38 |
| Magdalena | 561 | 531 | 349 | 203 | −38 | −42 |
| San Ramón | 558 | 459 | 302 | 186 | −46 | −38 |
| San Joaquín | 251 | 240 | 169 | 143 | −33 | −15 |
| Exaltación | 475 | 382 | 349 | 171 | −27 | −51 |
| Santa Ana | 332 | 338 | 371 | 282 | +12 | −24 |
| Reyes | 175 | 168 | 222 | 239 | +27 | +8 |
| San Borja | 8 | 13 | 41 | 6 | +412 | −85 |
| San Pedro | 400 | 268 | 181 | 101 | −55 | −44 |
| San Javier | 401 | 320 | 191 | 148 | −52 | −23 |
| Trinidad | 967 | 892 | 942 | 715 | −3 | −24 |
| Loreto | 544 | 430 | 269 | 241 | −51 | −10 |
| San Ignacio | 571 | 372 | 385 | 263 | −33 | −32 |
| Total | 6,407 | 5,446 | 4,500 | 3,050 | −30 | −32 |

*Source*: Boccolini and Jiménez, *Estadísticas de contribución indígena en Bolivia* (Lima: Biblioteca Andina, 1979), 248; *Revisita de la Provincia de Mojos 1854* (ANB RV Beni 1854, T.1), 13; *Matrícula jeneral del Beni para el Tribunal de Cuentas* (ANB RV Beni 1874 3), 106, 235, 295.

**Table 3.2** Absent tributaries (ausentes) in the four Mojo cantones, 1854–74

| *Cantón* | *1854* | *1867* | *1874* |
|---|---|---|---|
| Trinidad | 14 | 66 | 350 |
| Loreto | 31 | 21 | 45 |
| San Ignacio | 19 | 14 | 152 |
| San Javier | 29 | 2 | 75 |

*Source*: *Revisita de la Provincia de Mojos 1854* (ANB RV Beni 1854, T.1), 13; *Matrícula jeneral del Beni para el Tribunal de Cuentas* (ANB RV Beni 1874 3), 45, 60, 83, 194, 106.

We must note that the tributary populations of most cantones were already in decline before the beginning of the Bolivian rubber boom in 1864, and that in many the rate of decrease actually slowed after the rubber boom began. And in contrast to Arteche's report, San Pedro's poverty dated to well before the rubber boom. The prefecture of Santa Cruz considered the town to be "extremely ruined" in 1841, and used the colonial discourse of the miserable Indian to describe its people as "those unhappy natives" and "those poor Canichanas."[80] The judge José Manuel Montero echoed this view of San Pedro when he complained about having to move there to the minister of the interior in 1846: "If, in order to live in the best town in Mojos there is need of much resignation and philosophy, how much strength of spirit does one need, señor minister, to inhabit the most unhealthy, the most deserted, and the most unfortunate and miserable in the whole department?"[81] Despite this evidence of San Pedro's earlier decline, the increasing depopulation after 1867 is apparent in several other cantones. This trend is evident in the Baure cantones of Carmen and Baures, in the Itonama cantones of Huacaraje and Magdalena, in the Cayubaba cantón of Exaltación, and in the Movima cantón of Santa Ana. Of the four Mojo cantones, the only one which showed an increased rate of decline was Trinidad.

Thus most cantones (thirteen out of sixteen) had already begun losing tributaries before the rubber boom. How can we explain this? The officials who enumerated tributaries sometimes made mistakes. The total figure of 5,446 given for 1854 in table 3.1, for example, included 52 Indians who had reached the age of fifty, and who should have been exempt from tribute, making the actual total 5,394. But such recording errors cannot account for the majority of tributaries who disappeared. Neither did the cinchona trade take many Indians from the Mamoré or Iténez drainages, as its extraction was centered on the upper Beni River. Cinchona merchants there generally used Indian labor from the Caupolican region north of La Paz and exported the bark by mule to the Pacific coast at Arica.

Some sources mention outbreaks of smallpox and typhoid fever that must have contributed to the decline in population.[82] But the majority of vanished tributaries must have simply moved, either farther away from their cantonal seats so that they were not counted, or to different cantones, or even to Santa Cruz or Brazil.

The revisita of 1861 gives us a glimpse of Indian mobility on the eve of the rubber boom. The information provided for Exaltación makes it possible to determine the name, age, marital status, manzana, and present residence of fifty of its fifty-one ausentes. Twelve went to Brazil, with Pará specified as the destination of seven. Eleven had gone to Santa Cruz, eleven to Santa Ana, five to Reyes, four to Trinidad, and two to Magdalena. One each went to San Ignacio, San Pedro, San Joaquín, Baures, and Cochabamba. Of the forty-two whose marital status is given, twenty-seven were single, twelve were married, and three were widowers. Although the fifth manzana had more ausentes than the others, every manzana had a fair number, even the supposedly elite first manzana, so there is little evidence that parcialidad affiliation was related to the likelihood of migration or river travel. The ausentes' names also suggest that forty-one were Cayubabas, two were Canichanas, two were Cruceños, one was a Mojo, two were Brazilians, and one was a Mura Indian from Brazil. The ausentes ranged in age from eighteen to forty-one with an average age of twenty-seven. Their ages all fell between eighteen and fifty, the ages of required tribute payment, so escape from tribute may well have motivated migrants. In 1868, Franz Keller observed the prefect of Beni urging a local corregidor to collect the Indians' yearly tax and to not let them escape to Brazil.[83]

## Responses to the Rubber Boom

The former mission Indians of Beni came under great pressure in the second half of the nineteenth century. Liberal reforms, especially the abolition of redistribution and of most communal property, eroded the base of the old mission culture formed by Beni Indians in partnership with the Jesuits. The rubber boom created a demand for Beni Indian labor and lured an unprecedented number of carayana immigrants to the region, some of whom remained in or near the former mission towns, acquiring urban and rural properties for their stores and ranches.

The rubber boom posed challenges for both Indian elites and commoners. Cabildo leaders like Acuruza and the Ojopis, attempting to preserve some sort of social distinction in a new situation dominated by newly rich merchants and rubber patrones, acquired their own ranches with their own Indian commoner peons. Attempting to retain some of their traditional importance as mediators between carayana authorities and Indian commoners, they facilitated the carayanas' acquisition of labor for river transport and rubber gathering. They seldom openly contested the authority of their carayana superiors (though Franz Keller observed "ceaseless differences" between the cacique and the corregidor of Exaltación), seemingly preferring to join the new elite. Not all cabildo leaders learned to negotiate the new commercial culture that came to dominate Beni, however. Franz Keller wrote that two of the richest caciques, "the one at Exaltación, and the good old chieftain of Trinidad, had been swindled by unscrupulous rascals out of their whole fortune,—house and home, cattle and plate."[84]

Beni Indian commoners perhaps had become accustomed to cabildo leadership. In the century to century and a half that the cabildos existed before the liberal reforms of the 1840s and 1850s, cabildo leaders had acted more or less for the common good, managing communal lands and herds, distributing food and chicha, redistributing salt and iron, organizing religious fiestas, and fighting carayana encroachments in the late eighteenth and early nineteenth centuries. During the early Bolivian rubber boom, many Beni Indian commoners still obeyed their cabildos, going to work as paddlers or rubber tappers when ordered to. But many must have also realized that their cabildo leaders seldom did more than organize religious fiestas anymore.

Without their cabildos' effective support, Beni Indian commoners found several new ways to look after themselves. One way was to accommodate themselves as best they could to the new situation that followed the abolition of most communal property in 1856 by choosing the best patrón they could find. Some Baures from Concepción entered into a special relationship with Pastor Oyola and his wife, and though becoming rubber workers, there is evidence that they were able to preserve some of their community and cultural life in the Oyola barracas. And though whites perceived José Gregorio Acuruza as exploiting his fellow Cayubabas, his peons may have respected him as a patrón because he represented traditional authority, spoke their language, and came from the same culture. Banzer Toro de Añez relates that Acuruza is still well-remembered in Cayubaba tradition, and calls him both *gran cacique* and *taita* (father).[85]

## Relations with "Bárbaros" on the Llanos

A second strategy, which was employed by most of the former mission Indian groups, was to flee their towns or patrones and join the nearest group of independent Indians. The Cayubabas in particular cultivated close relationships with independent Indians like the Araras (or Caripunas), Pacahuaras, Chacobos, and Morés, which allowed them room to escape the labor demands of the rubber trade. Many Itonamas and some of the Baures of San Joaquin joined the Morés, and some Trinitario Mojos associated themselves with the Sirionós.

The Cayubabas had an interesting relationship with the Chacobos, a "docile savage horde" (in the words of Ignacio Araúz), who lived to the north and northwest of Exaltación. Even though Gibbon claimed that the Cayubabas and Chacobos were "constantly fighting" each other in 1852, this hostility did not last. In 1868, Franz Keller wrote that the Chacobos were peaceful and formerly visited Exaltación, but they more recently had retired to the interior and resented attempts by Cayubabas or Movimas to round up wild cattle in Chacobo territory. By 1869, Maccheti noted that a subgroup of Chacobos called Mayosas were "tame bárbaros and friends of the Cayubabas." He also baptized a thirty-year-old Chacobo man in Exaltación, who had a Cayubaba wife and two children. A Brazilian white served as this Chacobo man's *compadre*. According to Mathews, the Chacobos respected cacique Acuruza and his cattle, and became quite friendly and often came to Exaltación to trade while he was corregidor. The Franciscan missionary Fray José Cardús also noted friendly relations between the Cayubabas and Chacobos in 1883 and 1884. Of the latter Indians, he wrote: "Various of them are tame, and some are accustomed to leave for Exaltación every year during the fiesta, where they generally also provide themselves with some tools." Despite the Chacobos' participation in the Cayubaba religious festival, Cardús noted that they had lost whatever desire they might have had to become Christians upon seeing the "sad state" in which the Christian Indians lived. To the contrary, "various Christian Indians of Exaltación have fled from the town and hid among them [the Chacobos] in order to live in the same manner."[86]

Cardús found evidence that other former mission Indians were fleeing their towns to live with nearby unincorporated Indians. According to him, the Morés had accepted runaway Baures from San Joaquin and Itonamas from San Ramón, as well as an entire Itonama parcialidad that fled from Magdalena. From information provided to Cabrera, it would seem that the Morés also took in runaway Cayubabas, or at least accepted their presence. Cardús also observed that along the Río Cocharcas and other streams to the northeast of San Pedro lived "many families of savages" that the carayanas thought were Sirionós, who at times would work at the mill of the San Pedro hacendado José María Fresco. They hunted with the same long bows

*Figure 3.9.* Chacobo family. From Erland Nordenskiöld, *Indianer und Weisse in Nordostbolivien* (Stuttgart: Strecker und Schröder, 1923).

and arrows as the Sirionós, and had the same habit of never looking at one thing for long, always casting their gaze about like hunters or, as Cardús put it, like animals. As a student of Indian languages, Cardús noted that the Christian Guarayo Indians couldn't communicate with these Indians, but should have been able to if they were indeed Sirionós, as both groups spoke related Guaraní dialects. Instead, the missionary was able to distinguish words in their speech which were either Mojo or Baure, both Arawakan languages. He concluded that these "Sirionós" were really Mojos that had fled from Trinidad. The Sirionós might very well have accepted runaway Mojos, as they also incorporated both whites who had been kidnapped as children and the descendants of runaway black slaves.[87]

## New Settlements

Some Beni Indians came up with yet another response to the rubber boom by fleeing their towns and founding new settlements rather than joining independent Indians. Cardús related that several families of Movimas had left their town of Santa Ana to start a new settlement on the Arroyo Mato some fifteen to twenty leagues to the south-southwest, where feral cattle abounded. The settlement was very difficult to reach, but one person who saw it said that the Movimas had set up a cross there. The Trinitario Mojos followed the same course of action, but to an even greater extent. Crossing the Mamoré River, a large number of them moved to the pampas to the southwest of Trinidad to escape the authority of the prefect and of their own cabildo as well. There they settled at San Lázaro, Roma, Trinidacito, San Francisco, Rosario, Todos Santos, and San Lorenzo.[88] San Lázaro and Roma are barely mentioned in the sources and may well have been ephemeral. San Francisco, Trinidacito, and Todos Santos were new settlements, while Rosario and San Lorenzo were old estancias which received an influx of new residents. By 1886, San Lorenzo was set to become the site of the most dramatic resistance which any of the former mission Indians of Beni would offer against the demands of the rubber boom.

## Conclusion

The rubber boom was a time of change for the Mojos and the other former mission Indians. Trends such as migration, declining town populations, decreasing artisan production, disintegrating parcialidades, and intermarriage between carayanas and cabildo elites had begun before the rubber boom, but were now accentuated. And ex-mission Indians did die while engaged in rubber work, even if not to the extent that the existing historiography has assumed. Such trends, along with the dispersion of population away from town centers, made it harder to maintain traditional culture in the face of change.

There can be no doubt that the former mission Indians of Beni suffered exploitation by the carayanas during the rubber boom—but it is equally true that the Indians were not helpless victims. They were constantly exercising their agency, looking out for their own best interests and choosing the best strategies for survival that their changing circumstances offered. For cabildo elites, this could mean acquiring land and peons, or preserving their role as labor recruiters. And by continuing to organize religious ceremonies, cabildo elites positioned themselves as defenders of traditional culture.

Commoners had a particularly wide range of options before 1880. They could kill or maroon their patrones, flee to independent indigenous groups in the rubber zone, or move to Manaus. With such options, rubber work may even have been an attractive opportunity to escape the taxes, tribute, and continuing unpaid labor of the former mission towns. But after 1880, the rubber boom became more oppressive for commoners for several reasons. They lost easy access to Manaus when steamboats took over the Madeira River trade, and northeastern Brazilian drought refugees provided a new source of labor on the Madeira and the middle Amazon. Beni Indians had fewer chances to escape from, disobey, or dispose of their patrones as rubber extraction settlements proliferated on the rivers. At the same time, the Bolivian national government allied itself with the rubber interests just as the richest rubber area was discovered along the lower Beni River and its tributaries.

As the cabildos increasingly served the interests of the carayanas and of a few elite Indian families, Indian commoners took matters into their own hands in Beni just as they had in the rubber zone. Whether by fleeing to nearby independent Indians, founding new settlements of their own, or just choosing a patrón who showed some respect for their culture, the former mission Indians proved themselves capable of frustrating any carayana attempts to completely dominate them. How the Mojos reinterpreted traditional culture in their search to control their own lives will be seen in the next chapter.

CHAPTER FOUR

# *The Ventriloquist Messiah*

## Introduction and Sources

By the early 1880s, many Trinitario Mojos had moved to San Lorenzo and other settlements to the southwest of Trinidad to escape the labor demands brought about by the rubber boom. By 1886, if not before, an organized movement had emerged in San Lorenzo. In 1887, Mojos and carayanas fought each other for several months in what the prefect of Beni called a "caste war," as the latter tried to reassert control over their Mojo labor force.[1] Several carayanas and a great many Mojos died in this fighting, but the principal result was a massacre of Indian leaders by the carayanas and the suppression of one form of Mojo agency.

The sources that allow us to further explore the events of 1886–87 were all written by carayanas, and the three most detailed sources were written after the fighting subsided. On July 4, 1887, Miguel Antelo, the former subcorregidor of Rosario (one of the Mojo refugee settlements) finished a written account of his firsthand experience of the Mojo resistance movement and of the carayana response. On September 27, 1887, the ex-prefect of Beni, Daniel Suárez, completed a manifesto in which he defended his actions against the Mojos earlier in that year. And on January 1, 1888, the Jesuit father Gumercindo Gómez de Arteche recounted the events of 1887 that he and two other Jesuits had ascertained during their mission to Beni immediately after the end of the fighting. All three sources are biased, as Antelo and Suárez were concerned with defending their honor and their handling of the crisis of 1887, while the Jesuit Arteche saw the liberal elite of Trinidad as anticlerical and took the side of the Indians. Nevertheless, Arteche and Antelo both provide some insight into the Indian side of the story. And so, knowing that the sources we have may be limited and biased, let us explore what happened to the Trinitario Mojos in 1886 and 1887.

## Migration and Settlement

These sources offer differing interpretations of the Trinitario migration to the west of the Mamoré. Daniel Suárez claimed that these Mojos began to settle the southwestern pampas in the late 1850s, motivated solely by their hatred of the carayanas, since at that time "there weren't *enganchadores* [labor recruiters] because the rubber industry hadn't even been dreamed of." The priest of Trinidad, Cosme Damián Rivero, followed the Mojos to the pampas and persuaded some of them to return to the capital, or in Suárez's words, "to civilized life." Nevertheless, the Mojos' hatred of the increasing carayana population continued to simmer. With the coming of the rubber boom, "San Lorenzo became the refuge of all the swindlers who avoided the payment of their debts and the fulfillment of their contractual obligations." Moreover, according to Suárez, it was the Mojos' own immorality and drunkenness which prevented the return of those who went to the Madeira barracas.[2] The prefect clearly sided with the rubber interests in placing all of the blame on the Mojos' supposed moral faults and none on the carayanas or the rubber industry. Although there is no evidence as yet that Daniel Suárez was closely related to the brothers who comprised the firm of Suárez Hermanos, he was certainly their ally, as evidenced by the fact that the Suárez Hermanos press published his manifesto. In addition, the prefect himself had been a rubber patrón in 1883, when Júlio Pinkas noted that a Bolivian named "Daniel Soares" lived on the Río Madre de Dios with forty workers.[3]

The Jesuit Arteche recounted the Trinitario migration differently. To him, the Mojo movement to the southwestern pampas was a direct result of the rubber boom, and did not begin until the 1860s. Although failing to move en masse, many Trinitarios began to settle west of the Mamoré, as much to take advantage of rich new lands as to escape "the avarice of the carayanas," for which "almost all [who went to the Madeira] paid with their lives." But the prefects in Trinidad still needed Mojo crewmen for their merchant friends and sent officials to the west bank of the Mamoré to recruit crews, who punished those who resisted with whippings.[4] Though biased against the liberal authorities of Trinidad, Arteche's account seems much more believable than Suárez's.

Antelo's description of the situation west of the Mamoré in 1886 and 1887 provides us with the important information that not all of the Mojos who settled in that zone were commoners trying to avoid demands on their labor or the payment of their debts. He mentions one estancia named Llomomo that was owned by the cabildo intendant Santiago Tamo, and another called Asunta that belonged to the Sacristanes of Trinidad. Another estancia near Asunta belonged to Nicanor Cubene, who was called an *indio principal* by Suárez and Arteche. These elite Mojos may have spent more time in Trinidad than on their properties west of the Mamoré, but they were involved in the settlement of that area nonetheless. It is not inconceivable that they could have moved Mojo commoners from Trinidad to their ranches west of the Mamoré, much as José Gregorio Acuruza had transferred Cayubaba commoners from Exaltación to El Pilar. If this is what happened, then the settlement of the west bank of the Mamoré was the result of initiatives by both Trinitario commoners and Trinitario elites. In addition, by 1887 there were at least four carayanas living west of the Mamoré: Antelo himself, Wenseslao Añez, Nicomedes Bascopé, and Guillermo Ugarte.[5] Of these, Añez had been a rubber patrón on the northern rivers before 1883 and had transferred his operations to the Iténez system by 1884.[6]

The Trinitario movement west of the Mamoré was a series of migrations in which people were pushed by liberal reforms, labor demands associated with the rubber boom, and floods, and pulled by the prospect of rich, higher lands which were distant from the authorities. There are indications that one movement occurred from 1842 to 1845, soon after the Department of Beni was founded, and that another took place as a means to escape the flood of 1853.[7] It is certain, however, that the Trinitario refugees were established west of the Mamoré by 1883 or 1884, when Cardús noted that "not long ago almost all the Indians of Trinidad fled and took to the woods." Their reason for leaving, according to Cardús, was to escape from the authorities, who sent them by force to work in the rubber industry on the Madeira River. The missionary described their situation as follows:

> The Trinitario Indians now form four small towns, and are situated on the left bank of the Mamoré, near the Río Sécure. They live almost independently, with their own chacos, and some raising a few cattle; but at the same time they serve some hacendados established there, who have striven to build oratories to satisfy somewhat their religiosity. Some go to Trinidad to have their little ones baptized. Now the authorities try to temporize with them, because they have seen that said Indians are

exasperated, and resolved to defend themselves with their arms or to take to the woods entirely.[8]

By constructing oratories for the Mojos, the few hacendados in the region west of the Mamoré provided them with local ceremonial centers. These were both a way of keeping Mojos close by for use as labor and a tacit admission by the hacendados that they could not expect to exploit that labor without some contribution to the Mojos' ceremonial life. As carayanas expanded their rural holdings in the succeeding decades, they would follow this same formula of providing local religious centers in exchange for labor.

Living in the tropical lowlands made resistance through migration an especially attractive choice for the Mojos. The llanos were sparsely populated, and there were vast areas beyond carayana control. As we saw in the previous chapter, former mission Indians often had quite amicable relations with the unincorporated Indians who lived around them. The Mojos' methods of subsistence, slash-and-burn farming and cattle raising, were quite conducive to migration and actually required considerable movement between seasons and over the years. And while highland indigenous people used migration as a means of resistance, they rarely founded whole new communities in new territories, and it was surely easier for the Mojos to move away from their chacos (which they abandoned periodically anyway) than for highlanders to leave behind their well-defined, irrigated valley lands.

## Beginning of the Millenarian Movement

Whenever the Mojos settled in San Lorenzo and the other refugee towns, something new began to stir among them in 1886. According to Antelo, it was rumored among the Indians of San Lorenzo that on the night of All Saints' Day (November 1) of that year, a voice was heard in the middle of the plaza or on the road to the cemetery of that settlement. This voice told them that there would be another great flood that year that would completely cover Trinidad, turning it into a lagoon, and that to save themselves they should move to San Lorenzo with their families and cattle.[9] Whether or not such a voice was really heard, its message contained a millenarian vision. The present, corrupt world (Trinidad) would be destroyed by some natural disaster or act of God, not by any human revolution. Those who obeyed the voice and left Trinidad before the flood would survive in a purified world (San Lorenzo). Significantly, the voice chose to speak in the center of the plaza, the focus of Mojo ritual life, and on the road to the cemetery, where the Mojo ancestors resided. Moreover, the first two days of November are when the dead return to the world of the living throughout Catholic Latin America. The Mojos who heard the voice in San Lorenzo, or who believed the rumors of such a voice, could hardly have taken it for anything but the voice of a Mojo ancestor.

Antelo states that Mojos did indeed begin leaving Trinidad with their cattle, and that he informed the priest, David Egüez, and the prefect, Daniel Suárez. Suárez's policy was to let the people go where they might, but to appoint a new corregidor over them so they could not simply live as they pleased. However, he took no further action after appointing Antelo as subcorregidor of Rosario, the refugee settlement nearest to Trinidad, on April 29, 1886.[10] The date of Antelo's appointment is further evidence that large numbers of Trinitario Mojos had settled west of the Mamoré River before November 1, but news of the voice's prophecy may have given added impetus to the exodus. And much as the voice predicted, the floods of late 1886 and early 1887 were much higher than usual.

This inundation brought about a great scarcity in foodstuffs, and merchants like José María and Santiago Mendoza attempted to take advantage of the situation by selling rice and corn at inflated prices. Priced out of their staples, the Mojos survived by eating hearts of the motacú palm while clearing and planting for their next crop. Under such conditions, the Mojos could not afford to continue supplying crews for river transport. Those who went as crewmen complained that they suffered from insufficient rations of food, and those who were taken from their planted fields returned to find their corn all eaten by birds for lack of people to guard the crop.[11]

By the end of January 1887, the Mojo corregidor of Trinidad, Manuel Prudencio Semo, and subcorregidor Antelo were having trouble exercising their authority over the Trinitarios who had settled west of the Mamoré. The corregidor sent four cabildo officers to the west bank of that river to recruit paddlers, offering a wage of ten pesos for two months' service. One of the officers was the intendant Santiago Tamo, who owned the estancia of Llomomo, and another was Manuel Maleca. Antelo directed these officers to Trinidacito, incidentally hoping to use them to gather peons to help with his own harvest. The officers returned three days later with only one paddler, and recounted how two Mojos, José Cueva and Juan Masapuija, had spoken insolently to them.[12]

A little more than a week later on February 9 or 10, the same officers returned from Trinidad with two more cabildo members, one of whom was the cacique Juan José Noe. They brought a letter from the prefect asking for crewmen to man the boats of his foreign (*gringo*) friend Samuel Steir, a merchant who was bound for the Beni River. In this instance, the cabildo members could only recruit five peons who were harvesting Antelo's fields. Antelo wondered aloud where all of the Mojos had gone, and whether they had gone to San Lorenzo, but the cacique told him that he and the cabildo knew nothing. That night, Antelo and the cabildo members rode to the house of José Cueva. Antelo confronted Cueva over the latter's recent disobedience and offered him ten pesos to enlist his son as a crewman. Having waited in vain for Cueva's son, on Sunday, February 13, Antelo went again to Cueva's house, only to discover that the whole family had left for San Lorenzo the day before, carrying their Catholic religious image with them. The subcorregidor noticed several other Mojos preparing their oxcarts to go to San Lorenzo as well.[13]

On Monday, February 14, Antelo exhorted Juan Masapuija and the other peons who were working the fields to complete the harvest promptly, so that they could all go to Trinidad for Carnival. Masapuija said that he would not go to Trinidad, because he was going to San Lorenzo. Another Mojo named Mariano tearfully told Antelo that God was preaching in San Lorenzo, and that on Carnival Sunday God would show his body and put himself on the cross. All of the peons then returned to the harvest without giving more information. The subcorregidor then abandoned his plans of going to Trinidad, and told a Mojo from San Francisco, Asencio Tamo, to stay the night in Rosario, and to take him to San Lorenzo the next day, as Antelo did not know the way. Tamo agreed.[14]

## Antelo's Journey to San Lorenzo

Notwithstanding, Tamo disappeared and Antelo decided to go to San Lorenzo with his Trinitario cowboy Bacilio on the morning of Wednesday, February 16. On the way they caught up with Juan Masapuija, who begged Antelo for food for his hungry daughter. Antelo refused to help and continued on his way, reaching Llomomo, the estancia of the cabildo intendant Santiago Tamo, on the next day, where he found several more oxcarts bound for San Lorenzo.[15]

On Friday, February 18, Antelo left Llomomo after lunch and reached San Francisco at 3:00 p.m. There a mounted Trinitario named Lino Noe, who was driving thirty horses to San Lorenzo with his brother, invited Antelo to continue the journey with him. Traveling together, Antelo, the Noes and their horses, and the six oxcarts carrying their families and religious images reached a new estancia at midnight, where they encountered many other Mojos bound for San Lorenzo. Two of them greeted Antelo, including one of his own peons; this man and his wife gave Antelo roast chicken and fried plantains to eat. While Antelo drank coffee, the Mojos brought two of their images into the estancia's house and lit candles. They all stayed up until 2:00 a.m., praying and singing before the images.[16]

On the next day, most of the other Mojos left this estancia at 4:00 a.m., but Antelo and his companions did not leave until 10:00 a.m. By 4:30 p.m. they had arrived within a quarter league of San Lorenzo, where they met six Indians herding cattle. They greeted Antelo and told him that God was in San Lorenzo. While Antelo's Mojo companions took their two images down from their cart, cut poles, and raised the images on their shoulders, Bacilio went to talk to two twelve-year-old boys. He ran back to Antelo weeping, and said that it was true, and that the boys had told him: "It is God that is in San Lorenzo; he says, 'I am the Sacrament, I am Jesus the Nazarene,' and the deceased Tata Eduardo speaks, also calling all the people by first and last name as if he knows them all, and that Frutos Nosa, also deceased, speaks as well."[17] Frutos Nosa, we will recall, was the dominant figure in the indigenous politics of Trinidad in 1843–65 and had served as both corregidor and cacique. Tata Eduardo, because of his title meaning father, was a respected older male, and presumably a Trinidad cabildo member at some point in his life. Thus, in addition to Jesus Christ, two highly regarded Mojo leaders from the recent past were reported to be speaking to the people in San Lorenzo.

At five-thirty, Antelo reached the edge of San Lorenzo. It was a small settlement of three houses with new roofs of motacú palm thatch and eight others with old, straw-thatched roofs. There were also a few very deteriorated dwellings with hide roofs among the newer structures, left over from the old cattle-raising estancia. At the entrance to the settlement, macheteros were dancing, awaiting the arrival of the religious images. Nearer to the houses, a line of abadesas was waiting, each woman holding a lit candle, with an old man at their head. As Antelo rode up on horseback, each abadesa left the line and took his hand in greeting, after which the old man also came forward to greet the subcorregidor. His name was Andrés Guayocho.[18]

## Andrés Guayocho

Antelo states that Guayocho was eighty years old, and this is corroborated by the revisita of 1861, which gives Guayocho's age as fifty-four in that year. For someone so old, we know little of his life before his meeting with Antelo. According to the revisita of 1861, he was married to Teresa Tamo and had four children: thirteen-year-old Andrés, seven-year-old Patricio, four-year-old María Presentación, and one-year-old Feliciana. In 1861, Guayocho was living as an *agregado* to the Mojo population of the estancia of Rosario, meaning that he was attached to the settlement without being a community member. By 1867, however, Guayocho and his family had disappeared from Rosario and were not recorded anywhere in Trinidad or its estancias. The matrícula of 1874 provides no further information, since Guayocho was not listed as living anywhere in Beni. It may be that Guayocho had already settled on the isla of San Lorenzo by 1867, in which case he had lived among the Mojos west of the Mamoré for twenty-five years before his millenarian movement brought him to the attention of the authorities in 1886. Another possibility is that Guayocho left Beni to work as a boatman or rubber tapper in the 1860s and 1870s. One report that reached the national capital Sucre in 1887, somewhat confused regarding names, says that "Andrés Noco" (apparently Guayocho) had been a boatman to one Francisco Valderrama.[19]

Although his wife Teresa Tamo was a Mojo, several sources note that Guayocho himself was an Itonama. Rogers Becerra Casanovas states that his family name was originally Guayacho, and that it means "is from here" in the Itonama language. According to the same scholar, the Mojos changed the last name to Guayocho, which he says means "fighter" or "intemperate" in their language. Becerra is probably right about the ethnic origin of the name, for there are several Guayachos listed as living in the Itonama town of Magdalena in the padrón of 1854. The author of one secondary source claimed to have information that Andrés Guayocho was from Huacaraje rather than Magdalena, but did not cite his source. Wherever he was born, Andrés Guayocho had lived among Mojos from at least 1848, judging by the age of his eldest child. He may be descended from an earlier Itonama migration to the Mojo towns. Evidence for this hypothesis can be found in the revisita of 1861, which lists two Guayachos and one Guayocho as members of the manzana of Florida in Loreto.[20]

## Antelo's Account of San Lorenzo

When the images arrived, Guayocho and Antelo parted, and the latter went to take up lodging in a room of one of his peons. Bacilio brought in Antelo's baggage and told him that God was coming, so Antelo went into the adjoining room, where he found a Mojo whom he knew named Damazo Noe together with a sick old man and his wife in their hammocks. Antelo asked Noe where God was, and Noe took him back to the first room and showed him a door. When Antelo went through the door he found Andrés Guayocho, whom he greeted. Guayocho had two chairs brought out to the patio of his house and he and Antelo sat down facing the plaza, where about fifty Mojos soon gathered in front of them.[21]

Antelo began to address the crowd, telling them he had come to see them and San Lorenzo, to which everyone had traveled despite the flooded areas they had to traverse to get there. Then he began lecturing them on religion, telling them that Jesus had told his Apostles that he would not return until Judgment Day, and that he hadn't come back nor was he able to speak to them now. Antelo then advised Guayocho to think well about what he was doing so that the people would not say that he had deceived them. Then a man named Venancio Semo spoke in Mojo, which Antelo did not understand, and another man named Manuel Yujo immediately left the crowd and stood near Antelo. When Antelo turned his head to spit on the ground, Yujo shoved him on the chest so hard that he and the chair were thrown to the ground. Several more Mojos then jumped on Antelo, grabbing him by the arms, legs, and torso, and finally tied his arms behind his back, stood him up, and threatened him with a stick.[22] News of this treatment of a white by Indians first reached Sucre in a somewhat different version, again with confusion over names. In this account the corregidor of Rosario, Miguel Velasco (actually the subcorregidor Antelo), was brought before Andrés Noco (Guayocho), who called himself God. Andrés ordered the corregidor to kiss his feet. The corregidor replied "that such a God had previously been a Machoto [Itonama] Indian, and that he was not there to kiss the feet of Indians." A group of Indians then beat the corregidor, who escaped with great difficulty.[23]

In Antelo's own version, the Mojos took him and Bacilio, both tied up, to the atrium of the chapel of San Lorenzo. Antelo and the chapel's *doctrinero*, Manuel Muiva, both tried to convince the crowd to untie them, but the crowd took Antelo back to Guayocho's house without untying him. After Antelo pleaded to be untied again, Guayocho ordered it to be done. Antelo now found himself with a crowd of Mojos in Guayocho's house at about 9:00 p.m. Guayocho was seated on the ground against a closed door with a short whip in his hand; to one side of him was a small chair covered with a scrap of imperial cloth, and to the other side was a burning wick on the wall which provided the only light. The doors were closed, two men holding rolled-up whips stood against the wall, and the wick was suddenly extinguished. In the dark, Antelo heard a barely audible "false voice" speak to him in Spanish. The voice asked him why he had come, and Antelo only replied that he was sick and that he could not hear anything. After some moments' silence, Antelo heard Guayocho's voice, then José Cueva's, and then Guayocho's again, both men speaking in Mojo. The wick was lit again and Antelo got permission to leave the house in order to relieve himself, but with a guard of eight men to prevent him from fleeing. After Antelo relieved himself, he was returned to the house where, after a quarter hour, Guayocho appeared again. Antelo claims that the old man placed his hand on his shoulder and told him, "Now you're going to sleep with a woman." Antelo replied that he had not come to sleep with any woman, and Guayocho stepped back. A little later, the Mojos let Antelo go to sleep.[24]

The next day, February 20, was Carnival Sunday. Early in the morning the cabildo of San Lorenzo met in Guayocho's house, in the room next to Antelo's. The Mojos covered the door between the two rooms with a hide, but Antelo found a hole in the wall through which to spy on them. He saw Guayocho standing with his bastón or staff of authority in his hand. To the right was a table on which were many other staffs, some candles, and a black-looking statue which Antelo supposed to be a small Christ. Guayocho was talking and handing the other staffs one by one to the persons who approached him. Each Mojo who received a staff knelt and kissed its middle, and then stood up and kissed its head very reverently. After completing this ceremony, the Mojos removed the hide from the door, and Guayocho entered and invited Antelo to the chapel. They all went and prayed the rosary and the Christian doctrine, and sung salves, after which everyone returned to Guayocho's house. There, Guayocho spoke to the cabildo members, who all handed over their staffs, which were then returned to the table. Guayocho clearly acted as head of the cabildo, and later in Antelo's account, the intendant Santiago Tamo recognized Guayocho as its cacique.[25]

Antelo stayed inside Guayocho's house most of the day until he went outside in the afternoon and came across Guayocho bathing himself near the cemetery. The old man invited Antelo to bring his family and come live in San Lorenzo, and indicated some high ground where the subcorregidor could build his houses and some forest where he could make his chacos. Then Guayocho told Antelo: "Come here with us, secretly, slowly; don't tell the carayanas in town, don't go there, because if you tell the people, I'll have you killed with arrows, just so, *hombre*. We want to make our town here, people will come little by little, three years from now this town will be big. You for our corregidor, we don't want another carayana, and you will write to the bishop asking for Father Airan as our priest."[26] Antelo does not explain why Father Airan was Guayocho's choice for priest. Neither does he comment on his own reaction to Guayocho's combination of threats and flattery.

The next morning, Guayocho invited Antelo to the chapel once again, and after a cordial reception by the cabildo, they went there. Antelo was given the front seat and greeted by the doctrinero, who then took up a clay pot which contained burning incense. The Mojos then took down a small crucifix and placed it on a new chair, well upholstered with red cloth, which stood in front of the altar. They began to play a *caja* (a drum made from a tree trunk) and macheteros began to dance. They prayed the rosary and the Christian doctrine, and then returned to Guayocho's house, where Guayocho spoke to the people in Mojo. In the afternoon, Guayocho again invited Antelo to pray in the chapel. Soon after this, Guayocho came to Antelo's room and smeared the carayana's face with dust. Guayocho then took Antelo to meet the cabildo members, who were waiting to perform Carnival dances with a caja and a Bolivian flag. They invited the subcorregidor to dance around the plaza with themselves and Guayocho, which he did.[27]

The next day (February 22), Antelo again attended the Mojos' religious services. In the afternoon he went to Guayocho and said that he was going to return to Rosario the following day to harvest his corn and mill his sugarcane, and that he didn't want to go alone, but Guayocho wouldn't permit Bacilio to leave. Guayocho asked him to wait, after which the Indian leader gathered together his cabildo and a few women in his house. Antelo spoke to them about several points of religion, including baptism and Moses, to whom he said God spoke in the clear light of day in obvious contrast to Guayocho's ceremony in the darkened room. The carayana also told them that they could have neither a town nor a church without taking the government and the bishop into account, for those authorities would persecute them if they did not. This speech left the Indians in a pensive state of mind.[28]

At 6:00 p.m., Antelo asked Guayocho again to give him Bacilio so that he might leave the next day. Guayocho told him, "Don't go to town, don't tell them anything, the people say that they will help you with the work you have to do." Antelo responded that they already knew in Trinidad, because the cabildo officers had returned and were saying that it wasn't Jesus Christ who was speaking in San Lorenzo, but rather the Mojos' devils who told them not to go to work and not to pay their debts. The subcorregidor asked José Cueva if this was not so, to which the latter replied, "Yes, it is true." Antelo added,

"Who knows what they are thinking in town?" Someone said, "Now God won't come," and Guayocho responded, "But let's see, we'll all go inside."[29]

They all went in Guayocho's house and placed two chairs in the middle of the room where Antelo and Cueva sat down. The rest of the room was the same as on the night when Antelo first heard the "false voice." Speaking to Antelo of the small chair covered with imperial cloth, Cueva said, "In that little chair Jesus Christ is going to sit down, he's going to ask you why you have come, and you say to him 'I've come to see you with all of my people.'" It was around 8:00 p.m. when the burning wick was again suddenly snuffed out, and the "false voice" returned and asked Antelo why he had come, to which he responded as Cueva had directed. Then the carayana sensed that everyone fell to their knees in the dark, upon which they sang, "Santo Dios, Santo fuerte, Santo inmortal," an old prayer to the Trinity. Men and women then sang salves, after which the wick was relit. "That night many spoke in their language which, as I didn't understand it, I was ignorant of everything," as Antelo later recounted.[30]

The following day, Antelo went to ask Guayocho for Bacilio again, but Guayocho hid from him. Nevertheless, Bacilio appeared and saddled Antelo's horses, and told the subcorregidor that he was going to stay to fasten new straw on the chapel roof, that everyone was going to gather tacuara cane to renovate the chapel, and that he would send his son with Antelo. The boy never appeared, though, and Antelo asked José Cueva for help after spotting him going to gather tacuara. Cueva arranged for Antelo to travel with two oxcarts going to San Francisco, which could pass the flooded arroyos. After agreeing to lend Bacilio's mount to an Indian returning to Rosario, Antelo and the other travelers left San Lorenzo. Two leagues from town, they encountered thirty mounted Indians with twenty-five teams of yoked oxen making a path through the high grass so that the tacuara could be transported to San Lorenzo. At 5:00 p.m. the travelers arrived at San Francisco, where all the houses were full of people, mostly women. The next day, Antelo ran into his peon Asencio Tamo in San Francisco, who again promised to travel with the subcorregidor, and again disappeared. Antelo continued on, heading for Trinidad despite Guayocho's wishes, where he arrived on February 28 or 29.[31]

## Analysis of the Millenarian Movement

Antelo's account is the only firsthand record we have of the Mojo movement that later came to be called the Guayochería after Andrés Guayocho.

We must now turn our attention to what Antelo and a few other sources can tell us about the ideology of the Guayochería. By examining the Native American and Christian influences on the Mojo belief system, and by placing Guayocho's movement in pan-American and even world perspective, we can deduce what was likely true about the movement's ideology, and what it meant.

Daniel Suárez called Guayocho an Itonama *Choquigua*, "that is, priest, prophet, the last vestige of the old religious beliefs and superstitions of this race."[32] Alcides d'Orbigny recorded *Chukiva* as the name of an Itonama spirit, which he called evil because of his Christian sources. He contrasted the Itonamas, who conserved a great deal of their customs and were only superficially Catholic, with the Mojos, who were "the most decided Catholics of the province," though not without their own superstitions.[33] The Itonamas experienced the shortest missionary presence among all Beni Indians–only forty-seven years—so they are more likely to have retained indigenous religious beliefs and practices than the Mojos, who had the Jesuits among them for ninety-three years. The writer and newspaper editor Juan Coimbra, who participated in the rubber boom, wrote that the Itonamas, despite Christian indoctrination, still recognized a "superior and supernatural being" or "dark spirit" called Choquigua which was "the inexorable arbiter of the destiny of man."[34] Rafael Arteaga Terrazas, an early historian of the Bolivian rubber boom, observed that the ability to communicate with the spirits of the dead was called *hacer choquigua* (which could be translated as "to play choquigua" or "to pretend to be choquigua").[35] Coimbra used a more correct term for Andrés Guayocho, calling him a *choquigüero* (someone able to contact spirits) rather than Choquigua (a spirit), and described the séances of such choquigüeros as follows: "In the middle of a sort of black mass which is celebrated in the dark, in the greatest mystery, with the trembling of exorcism and the noise of manipulations, [the Choquigua] is invoked by the witches [*brujos*] who 'feel' and interpret it."[36]

Guayocho's performances were also done in the dark, and one supernatural being he interpreted was Jesus Christ, who came and sat in the small chair beside him. Antelo also implied that two deceased Mojos, Tata Eduardo and Frutos Nosa, were speaking through Guayocho. The first report to reach Sucre added that "Andrés Noco" (Guayocho) had "the jaguars and caimans of those remote regions as obedient soldiers." Jaguars and caimans represented the two most powerful spirits of the pre-mission Mojos, and they reappeared in a twentieth-century Mojo millenarian movement influenced by the Guayochería, in which the terrestrial paradise (the Loma

Santa, or "holy hill") was thought be guarded by jaguars and caimans, which would let the Mojos in and keep the carayanas out.[37] Guayocho thus manifested their most important supernaturals to his Mojo audience: Jesus Christ, their own recent ancestors, and jaguars and caimans.

It is apparent that Guayocho was a shaman, an individually inspired religious practitioner who was able to directly contact spirits. His title of choquigua or choquigüero suggests as much, as does his technique of creating a "false voice" to represent the speech of supernatural beings, which was labeled "ventriloquism" by Daniel Suárez.[38] Guayocho was not entirely unique for his time or region, as other Amazonian shamans led millenarian movements against the disruption of the rubber boom in the late nineteenth century.[39] In Bolivia itself, there was another millenarian movement among the Ava Guaraní (Chiriguano) Indians of the Andean foothills in the early 1890s, whose leader was a shaman who claimed supernatural connections.[40]

Broadly speaking, there are three widespread beliefs among world cultures regarding how a person can have contact with spirits or other supernatural beings. In one method, a spirit comes and enters the body of a person who is then possessed. According to Erika Bourgignon's survey of 488 societies around the world, belief in spirit possession was widespread in the Pacific, Africa, Afro-America, Europe, the Middle East, Sri Lanka, and Indonesia. The lowest incidence of belief in spirit possession was among the native peoples of North and South America.[41]

The low incidence of possession among Native Americans is related to the prevalence of other techniques of contacting spirits. In one method, also common in much of Asia, the shaman's soul leaves his or her body and goes to meet the spirits in their own world.[42] A flight to the spirit world was an individual experience that could not be manifested to others in the way that spirit possession could. In cultures that do not believe spirits can occupy human bodies, an alternative was to bring the spirits down alongside the shaman and to have them speak through ventriloquism. In a survey of Siberian shamanism, Anna-Leena Siikala identifies the ventriloquist séance as a third method of contacting spirits, along with possession and soul flight.[43]

Shamanic ventriloquism appears infrequently in historical and ethnographic sources. Of eighteen identifiable cases, thirteen are from areas where flight to the spirit world was common (the Americas and northern Asia), and twelve of these are from the Americas.[44] The bulk of the evidence demonstrates that in areas like the Americas where spirit possession was rare, shamans could use ventriloquism to manifest the presence of spirits to their audiences. Guayocho was not the only leader of a late-nineteenth-century indigenous movement to use ventriloquism. The leaders of the Maya caste wars in both Yucatán and Chiapas in Mexico utilized the same technique.[45]

Despite the difference between ventriloquism and spirit possession, both are forms of discursive agency, and the practitioners of both can gain power and authority by shifting their identity from devotee to powerful supernatural being.[46] By interpreting the voices of Jesus Christ, Frutos Nosa, Tata Eduardo, and perhaps jaguar and caiman spirits, Guayocho augmented his own power and authority. He would never have spoken in the voice of someone *less* powerful or authoritative than himself. As an exercise in individual agency, Guayocho's performances allowed him to advance from agregado to cacique, and to vie with the traditional elite for authority over the Trinitarios. Guayocho's religious movement was also an exercise in group agency, which aimed to renegotiate the shifting power relations between Mojos and carayanas.

Guayocho's religious movement was clearly something new. Although extant scholarship has assumed a precontact propensity among the Mojos for millenarianism and migration (in the manner of the Tupí-Guaraní search for a "Land-without-Evil"), and continuity from these beliefs through Guayocho's movement to the Loma Santa movements, there is little proof of such continuity.[47] Nor are claims that Guayocho's shamanism was descended from the practices of pre-mission Mojo shamans convincing. Guayocho belonged to a strong Itonama shamanic tradition that was still vital in the late nineteenth century. Mojo and Itonama shamanism may have been very similar in the seventeenth century, but Itonama shamans retained more of their practices into the nineteenth century, which is why the Mojos regarded Guayocho as possessing supernatural power.[48] At first it may seem strange that the Mojos accepted an Itonama as their leader in San Lorenzo, even accounting for Guayocho's long association with Mojos. In reality, peoples around the world often attribute greater supernatural power to an ethnic other, and Itonamas had a reputation as choquigüeros throughout Beni. It was also quite common for millenarian movements to coalesce around a charismatic outsider. The Mojos themselves followed another outsider (a Guarayo) in a Loma Santa episode in 1959.[49]

Daniel Suárez claimed that Guayocho tried to take the congregation from the priest of Trinidad, and wanted to be worshipped as the incarnation of God, "setting himself up in the chapel of San Lorenzo alongside the images of the saints, as a true object of adoration." The prefect summarized that "in a word that insurrection was against the state religion, against civilization, and against morality."[50] Through such statements, Suárez tried to make it seem that Guayocho and his Mojo followers were abandoning civilization and rebelling against the state and its official religion, obviously as an excuse to use force in order to recapture their labor for the rubber industry. Despite some innovations, the movement in fact showed little evidence of either political or religious rebellion. We have already noted that Trinidad was supposed to be destroyed by a flood, not by any human insurrection. Guayocho claimed that God spoke alongside him, not that he was God, and his séances took place in his house rather than in the chapel. Within the chapel and everywhere else but Guayocho's house, the Mojos did nothing new or unorthodox with their religion, and Guayocho even requested that a priest be sent to them.

The Mojos of San Lorenzo also respected established forms of political authority, only rejecting their contemporary representatives in Trinidad. They carried a Bolivian flag in their Carnival dances, and set up a cabildo with the same symbols of authority, canes of office and whips, that were used in the former mission towns. The one very important political innovation of the Trinitario refugees, and a prime example of their agency, is the fact that they created a *new* cabildo in San Lorenzo to replace the authority of the Trinidad cabildo. Given the millenarian aspect of the movement, the San Lorenzo cabildo was set up to be a successor rather than a rival to the one in Trinidad. Guayocho's importance to the refugees was to give supernatural sanction to their abandonment of Trinidad and creation of a new cabildo. The San Lorenzo cabildo was the first new cabildo to be founded since the eighteenth century, and the first to be created exclusively by indigenous people.

Finally, we must consider the possibility that Guayocho was not such an important leader as Antelo and Suárez made him out to be. Arteche says that few of the Mojos actually believed that God was speaking in Guayocho's house, but that they respected Guayocho for his message: that God felt sorry for how the carayanas had treated them and had decided to take them to a place far away where they could live and practice their religion in peace.[51] Some Bolivian historians have even put forth the thesis that Guayocho's leadership and his millenarian ideology were both inventions of the carayana authorities, who used them as an excuse to employ violence to regain control of Mojo labor.[52] Such an argument is unlikely to be true, since Antelo and Suárez both agree that Guayocho was a leader and a shaman, even though the two carayanas became enemies, frequently disagreeing about other points and attempting to blame each other for the crisis of 1887.

## The Carayana Reaction: The First Expedition

Miguel Antelo arrived at Trinidad's river port on February 28 or 29 at about 5:30 p.m. Word that he had been mistreated by the Indians in San Lorenzo was already circulating, and many of his friends met him at the port to ask questions. When he arrived at his house in town, other carayanas as well as Mojos gathered to hear his story. The next morning around nine o'clock a soldier arrived to escort him to the prefecture. As Antelo passed through the barracks on the ground floor, he noticed that people were gathering there. On the second floor, he met the prefect, and began to tell his story. In Antelo's account, Suárez cut him off, saying that he had already been told what had happened the night before, when the priest David Egüez entered.[53]

Antelo says that the prefect told the priest, "I'm going to order that the delinquents and all the people there be brought for punishment, and that the chapel be burned." Egüez protested that the chapel was blessed and that some of the Mojos living west of the Mamoré had owned land there for many years, and added that he would tell the bishop. Antelo claimed that Suárez quickly decided to send a military expedition against San Lorenzo after consulting with five or six friends, of whom two were opposed to such an action. Suárez claimed that Antelo, "having been badly treated and injured" in San Lorenzo, gave him an account which "was dictated by resentment which cried out for vengeance." Antelo was probably much more upset at his treatment in San Lorenzo than he admitted in his own account, which he wrote with the purpose of escaping any blame for subsequent events, and his first report must have been the source of the story of the corregidor and Andrés Noco that appeared in the Sucre press. At any rate, we know that the prefect was sufficiently concerned to cancel a planned inspection of the outlying cantones of Beni. As Antelo left the prefecture with Egüez, he noted that more people had gathered in the barracks to join the expedition. In the barracks, Sergeant Benigno Velasco told Antelo that he would have to go on the expedition with them to show them the way to San Lorenzo.[54]

On March 8, the first part of the expedition, composed of a troop of collas, crossed the Mamoré. In the environs of Asunta, the estancia of the Trinitario Sacristanes, they requisitioned almost all the horses without paying their owners, telling them that Antelo would pay them later. Antelo says that Velasco and others remained in Trinidad, spending the greater part of the day drinking in the barracks or in Antelo's room. At 10:00 p.m., Antelo went to look for Velasco to ask when they would depart and found him in the church atrium talking with Egüez. There, the priest told the sergeant, "Even if you have an order, be it written or verbal, to set things on fire, don't comply with it, because you'll be responsible; there are old landowners [west of the Mamoré]." Antelo went to the prefect and told him it would be better if Egüez went with them, but Daniel Suárez denied the request, saying that the vicar would disobey orders. When Velasco continued drinking, he told Antelo in a loud voice, "We're going to avenge what these Indian rogues have done to you; we'll tie them up and give them 200 lashes and bring them back tied up; I carry written orders and verbal ones, too." The Indian crew that was waiting to take the carayana expeditionaries across the Mamoré heard Velasco's boast, and Antelo, annoyed, told the sergeant and his men, "Now this, sirs, is a joke; see how the crewmen hear everything." Daniel Suárez later denied giving Velasco any verbal or written order to take the offensive. At 3:00 a.m., Antelo finally heard the bugle call that signaled that it was time to depart.[55]

On March 9, Velasco's part of the expedition reached Asunta on the west bank of the Mamoré. Around 10:00 a.m., Velasco ordered his men to requisition horses and a yoke of oxen from the surrounding estancias, including that of Nicanor Cubene. They only found three mares, so Velasco and his men marched to Antelo's house on foot while two men searched the estancias on two of the mares and Antelo ordered the third returned to its owner. At 5:00 p.m. they arrived at Antelo's house at the port of Rosario, and Velasco and his soldiers passed on to Rosario to join up with the colla troop.[56] The complete expedition included thirty carayana soldiers and a number of Mojo guides. Arteche later gathered the information that almost all of the soldiers were given to drunkenness and were publicly cohabiting with women.[57] Whether or not this is true, Antelo's report reveals that they certainly lacked discipline.

On March 10, Velasco informed Antelo that all was ready, but Antelo waited for his baggage to arrive from Trinidad. When it arrived at 6:00 p.m., he saddled his horse and went to the soldiers' camp. Soon after, seven horses arrived, sent by Nicanor Vaca, who had been named *alcalde de campo* (a post equivalent to subcorregidor) by the Trinidad authorities. Antelo wondered whether he was still subcorregidor, since he had neither been informed of Vaca's appointment nor charged with requisitioning horses himself.[58]

The following day, the expedition left Rosario at four o'clock in the morning, and reached Llomomo, the estancia of the intendant Santiago Tamo, at noon. The estancia was named for a nearby *yomomo*, a Mojo term for a body of still water completely covered with floating plants.

Upon sighting the estancia's house, several soldiers began to race to it to see who could get there first. The race ended for some when they tried to cross the yomomo and fell through the aquatic plants into the water, thoroughly soaking their cartridges. Arriving at the house, the other soldiers found the doors closed and no one there. They went in, looked around, and found two jugs of chicha, which they drank. When Antelo arrived, they had begun to drink some sugarcane juice which they had found as well. Antelo says that he reproached the soldiers for helping themselves to Tamo's goods, especially since the intendant was traveling with them and might become annoyed.[59]

After eating what they could find in Tamo's house for lunch, the expeditionaries left Llomomo around 2:00 p.m. and marched to San Francisco, where they arrived at five-thirty in the afternoon. The settlement was completely empty of people and goods, except for some old trunks and hatchets, and the houses were closed up. As night was falling, the soldiers unloaded their supplies and Velasco ordered them all to gather in the atrium of the chapel, except for sentinels whom he placed at the plaza's corners with fires. A soldier opened the chapel doors, and the rest followed him in. Antelo noted that all of the religious images were gone, and told Velasco, "Now this is a serious thing . . . there could be a strong resistance, now there are a lot of people in San Lorenzo." Velasco decided to send a letter asking for help from Trinidad, and Antelo offered to take the letter along with Tamo and two crewmen.[60] Arteche says that the troops lodged inside the chapel that night, and that they cooked their dinner in the sacristy.[61] With the entire town empty, such treatment of the chapel was hardly necessary.

Early on the morning of March 12, Velasco dictated the letter to Rosendo Mejias, asking for arms, munitions, and more men, and advising Trinidad that his troops were at the point of mutiny. In response to this feeling among the men, Velasco told them that anyone could leave if he wanted to, but no one could take a horse with him. When the letter was ready, Velasco ordered Tamo to appear so he and Antelo could return to Trinidad. Tamo said that he didn't want to cause a commotion in Trinidad, and that he would go to San Lorenzo to bring back the cacique Andrés Guayocho to talk with them. After some discussion, the carayanas agreed to the intendant's plan, and told him, "Go then four of you, bring your relatives, and tell Guayocho to come with his cabildo and that we won't do anything to him; be here tomorrow at noon, we'll wait for you." Four Mojos then left on horseback, among them Santiago Tamo and his fellow cabildo member Manuel Maleca.[62]

The next day, the carayana troops got up early in the morning, saddled their horses, and prepared to march on San Lorenzo instead of waiting until noon as they had promised. Antelo told Velasco that they would have to cross a deep arroyo and that they would have to use canoes to cross, but the commander took off his clothes, walked into the water, and found it only came up to his waist. Antelo was impressed with how rapidly the water level had dropped. As the troops entered the arroyo, Velasco commented how nice the water was, and Antelo said they should all bathe to be fresh for their march.[63]

Their two carts crossed the arroyo and continued on, followed closely by the troops. After marching a half league or less, they followed the trail into the woods until they were stopped by trees which had been felled across the path. Suddenly, the Mojos fired on the expedition from the woods, using bows and arrows and "twenty-five old shotguns that were at the estancias to defend them from jaguars," according to Arteche. The Jesuit believed that the Mojos attacked out of rage at the profanation of the chapel of San Francisco. The expeditionaries were caught completely off guard, traveling "as if they were hunting rabbits" rather than preparing to engage an enemy. In Antelo's version, they were walking their horses with their unloaded rifles tied on the animals, which bolted for the woods, some frightened, others wounded, at the first sound of the Mojos' shotguns. Other accounts state that the carayanas' weapons were all on the oxcarts which traveled at the head of the column. Some carayanas died at the ambush site, and the rest fled. Antelo describes his own comportment thus: "I dropped the reins of my horse to seize four horses that left with guns, at which time nine soldiers took off like a shot in escape, some without hats, and so yours truly did the same." Mejias tried to make a stand at the chapel of San Francisco with a soldier named Saucedo and some others, but the rest abandoned him and continued their flight. The pursuing Mojos killed those at the chapel, and the other soldiers that they could catch on the road. Of thirty expeditionaries, twenty-one were killed, mostly collas, and mostly by arrows. A Santa Cruz newspaper later identified some of the dead as Velasco, Mejias, the comisario Marcos Vargas, a certain Ortíz, "many collas," and it was thought, one Antenor Valverde. In addition to inflicting these deaths, the Mojos made off with eleven rifles and some munitions. The fate of Santiago Tamo and his companions is unrecorded.[64]

Antelo and the nine soldiers fled all day with their horses and guns, arriving at his house at the port of Rosario around 8:00 p.m., after a journey of fourteen leagues. As soon as they reached the port, eight of the soldiers embarked for Trinidad. Antelo arrived at his house with one soldier a bit later, as their horses were limping. The subcorregidor found two families at his farm, one of his mayordomo, and the other of his cowboy Bacilio, who had returned from San Lorenzo saying that he didn't want to go back. The next day, Antelo and Bacilio rode to Asunta, and the cowboy paddled the carayana to Trinidad in a canoe. Antelo left most of his possessions as well as the two families behind and was unable to obtain a crew in Trinidad to go back for them.[65]

## Interim

The defeat of Velasco's expedition and the death of twenty-one of its men was not the only troubling news to reach the carayana authorities in Trinidad. Around the same time, probably on March 11, an Indian crew killed the Cochabambino merchant Mariano Valderrama and his son while on a river voyage. The carayanas then arrested several Mojos, including a servant of Francisco Valderrama, Mariano's brother and Guayocho's presumed former patrón. According to Arteche, this servant denied any part in the murders, but then retracted his statement and confessed after the carayanas had given him two hundred lashes. Francisco Valderrama then ran to the carayana tribunal to tell them that his servant had been with him on the day of Mariano's death, and couldn't have had any role in it.[66]

Immediately after the remnants of Velasco's expedition reached Trinidad, the prefect began to plan a second one. He requested arms and people from Trinidad and from the other cantones in the Department of Beni. In the departmental capital, Rómulo Suárez donated all of his rifles, amounting to some fifty, to the cause of suppressing Guayocho's movement. The commissary, Felipe Laime, and the police intendant, Napoleón Vaca, began to take a roll call every night and soon found that some of the citizens of Trinidad were fleeing while others expressed their opposition to a second expedition. The latter were then targeted for recruitment by the authorities, according to Antelo. The prefect called for a committee, to which the priest Egüez and the citizens Marcelino Marañón and Rosendo Gutiérrez were opposed. Then the authorities published a proclamation that anyone who left Trinidad without a special police passport would have to pay a two-hundred-peso fine and would be considered to be in league with the rebels of San Lorenzo. Next, the authorities called on all the citizens to form a National Guard, and named Rómulo Suárez as its chief. Finally, another meeting took place at the prefecture at which several important carayanas decided to form a Committee of War, which Antelo says included Daniel Suárez as president, and Rómulo Suárez, Nicolás Suárez, Manuel Santos Moreno, Isaac Michel (president of the Municipal Council of Trinidad), Carmelo Fernández (vice president of the same council), and Ignacio Aguilera as members. Arteche called it a Committee of Public Safety, or Comité de Salud Pública, to rhetorically link the liberal elite of Trinidad to the excesses committed by the French Revolutionary committee of the same name. Whatever its name, the committee set about raising money, and Antelo says that it forced other citizens to contribute, and imprisoned one without food until he paid.[67]

After the destruction of Velasco's expedition, Guayocho seems to have set about gathering all of the Mojos west of the Mamoré into his own group. On March 21, the civil and religious authorities of Trinidad sent four Indians on a peace mission to the west bank of the Mamoré. At the home of Pedro and Lino Noe, the emissaries encountered twenty Mojos from San Lorenzo, who wanted to take two of them with them. In the end they only took one, named Manuel José Cueva. The three remaining emissaries continued on and met Guayocho and his followers, and "found them very furious, submitting them somewhat to their dominations." Meanwhile, the party of twenty Lorenzanos went to Antelo's house at Rosario, took out his trunks and other household goods, burned his buildings, and carried off his goods, carts, oxen, horses, and the families that were there along with Antelo's mayordomo. The three emissaries encountered this party with their spoils while they were returning to Trinidad on March 25.[68]

On March 26, a wealthy Italian resident of Loreto, José María Ferrera, wrote a letter to his friend and relative in Santa Cruz, Ricardo Chávez, the former patrón of Carapanatuba on the Madeira River. Ferrera related how the Indians of San Lorenzo had rebelled, mistreated Antelo, and killed twenty-one of the thirty soldiers in Velasco's expedition. He described how the prefect had asked Loreto to send armed and mounted men, and how the important citizens had each contributed ten pesos, but had not sent arms as Loreto was closer to the rebels than Trinidad. Ferrera wrote that the river commerce to Cochabamba and Santa Cruz had been interrupted, and that it was said that the authorities had captured an Indian in Trinidad (probably Francisco Valderrama's servant) who told them that there were many rebels, well armed with bows and guns and with good horses. The captive was also alleged to have said that the Mojos had planned their rebellion for a long time, and that they now proposed "to put an end to the white race of Trinidad." Yet Daniel Suárez was not as eager as Ferrera to tell the outside world what was going on in Beni.

On the same day, Suárez wrote a letter to Manuel Argandoña, the new prefect of Chuquisaca, congratulating him on his nomination and proposing future political cooperation, but offering no hint that he considered himself or his race to be threatened with extermination.[69]

Suárez's unwillingness to inform the national government of his actions during the Guayochería became one of the most controversial aspects of his term of office. In his own defense, he claimed to have informed the minister of the treasury on March 28 and the minister of war on March 29 about Antelo's misadventures and the decision to send Velasco's expedition, and to have requested military aid from the latter minister. Yet the only communiqué to be found in the National Archive now is a note from Suárez to the minister of government in which the prefect relates how he published a certain new national regulation for the police, and sent it to the other provinces of Beni. Nowhere does he even hint at any problems with the Mojos.[70]

Sometime thereafter, the Mojos of Trinidacito joined Guayocho's cause and went to sack the property of Wenseslao Añez, where the Alcalde del Campo, Nicanor Vaca, was living. Vaca received word in advance and escaped to Trinidad in his shirtsleeves, and Mojos from Trinidacito and San Lorenzo carried off the horses of Añez and of Nicomedes Bascopé. Guayocho went to San Francisco in person to take away its people, while the people of Trinidacito sent him a party headed by Lino Noe.[71]

## The Second Expedition

Preparations for the second expedition continued. Arteche says that "no one lent themselves to go in it, and in order to be able to assemble the 100 carayanas and fifty Canichanas that finally went, it was necessary to threaten them liberally with 500 lashes." In Antelo's version, only forty Canichanas went. Two days before the expedition left, Antelo was informed that they would try to take him along, and so he fled Trinidad for the countryside. The second force left Trinidad on May 6, commanded by Nemesio Saavedra, with Félix Montenegro as second-in-command. Daniel Suárez claimed to have written to the minister of war on May 7 about Benigno Velasco's death, and to the minister of government on the same day about the death of Marcos Vargas. Once again, no such letters are to be found at present in the National Archive.[72]

None of the chroniclers of the Guayochería went on the second expedition against San Lorenzo, but Antelo and Arteche provide some information about what happened. The expeditionaries easily reached San Lorenzo, from which the Mojos fled. The latter took few of their possessions with them, and left the rest in the chapel, hoping it would be respected. Some, mostly women with babies, hid in the tall grass and died when the expeditionaries set fire to it. The soldiers caught up with some of the Mojos just as they were about to enter the woods with their carts. There they killed four children and an old man named Eugenio Noe, the father of Cacique Juan José Noe of Trinidad. They also killed an old woman after stripping her of her clothes, and raped several other women. The soldiers took ten prisoners, including Andrés Guayocho and Juan Masapuija. After whipping these captives in San Lorenzo, the carayanas marched them to Rosario. There they once again whipped Guayocho mercilessly. The old man, having lost a lot of blood, then asked for a drink of water. In response, Montenegro kicked Guayocho in the mouth and caused him to tumble to the ground. The carayanas then began shooting the choquigüero, killing him slowly by first aiming at his legs "and less vital parts." They also shot Juan Masapuija to death and whipped the eight remaining prisoners in Rosario. Of the eight, one died after receiving more than six hundred lashes, and another after being given nine hundred. The Canichanas spent a week searching for Mojos around San Lorenzo, and killed all those they found, both adults and children. Their willingness to help in this task suggests that the old enmity between Canichanas and Trinitarios continued to simmer. Some of the members of the second expedition returned to Trinidad on June 12.[73]

Antelo says that Saavedra and his men sacked, robbed, and filled sixty carts with the trunks, tools, and other possessions of the Mojos. They burned houses, chapels, chacos, and destroyed orange groves. They also stole two thousand head of horses, of which only one hundred reached Trinidad, where they were sold for a very low price at a public auction at which Rómulo Suárez was the only bidder. Saavedra and his men took the rest of the horses, the sixty laden carts, and two more carts carrying captive women and children to San Ignacio. There they divided the spoils, including the oxen that had pulled the carts, among themselves. Both the carayanas and the Canichanas enriched themselves with booty. In addition, the expeditionaries still had not returned the carts, oxen, and arms that they had been lent in Trinidad when Antelo wrote his report on July 4.[74]

## Repression in Trinidad

As he dispatched the second expedition to the savanna, Suárez and the other carayana leaders turned their attention to the Mojos of Trinidad.

The carayana authorities became convinced that the principal Mojos of Trinidad, including cabildo members, were in league with Guayocho and his followers. Perhaps their suspicions came from the fact that elite Mojos had estancias on the west bank of the Mamoré, or that the cabildo intendant had gone to see Guayocho just before the attack on Velasco's expedition. At any rate, they believed that the Mojos who had remained in Trinidad and its environs intended to burn the capital and murder all of them, and that Nicanor Cubene was behind the plot and had been behind the murder of Valderrama.[75]

Consequently, when a large number of Mojos filled the church on May 9 for the Ascension Day mass, the carayanas surrounded the building and took them prisoner. In Arteche's words, "those who didn't hear mass captured those who did hear it, and the church, which in better times was the asylum of malefactors, served those liberals then as a trap to take Christians." Of the 140 Mojos whom Daniel Suárez says were taken prisoner, the carayanas took all sixty men they had captured to the barracks along with some of the women. There, headed by the police intendant Napoleón Vaca, they gave the Mojo men each five hundred to six hundred lashes, and the women each two hundred to three hundred, over three days so that they would confess the truth. One carayana later told Arteche candidly that the Mojos were whipped so that they would confess what the carayanas wanted to hear. Antelo called the carayanas responsible for the whippings a "Tribunal of Lashes," and said that it included the municipal council president Isaac Michel, the vice president Carmelo Ignacio Fernández, Rómulo Suárez, Ignacio Aguilera, Gumercindo Arriaza, Nicolás Suárez, and two interpreters. Damiana Semo, the wife of Cacique Juan José Noe and daughter-in-law of the Eugenio Noe who was killed in San Lorenzo, told Arteche that ten Mojos had died from the whippings. These victims included the cacique, the corregidor and sacristán mayor Manuel Prudencio Semo, Nicanor Cubene, Cubene's wife Nicolasa Nosa, and six other men. The carayanas reportedly asked Nosa to testify against her husband. She replied that she knew nothing against him, but would never say so even if she did. The carayanas then whipped her, and after more than two hundred lashes, she died.[76]

The carayanas' arrest and killing of such prominent Mojos understandably became one of the most controversial of the events of 1887 in Beni. The corregidor and cacique were, after all, important members of the local government, and yet they were killed by other representatives of the same government. The carayanas overlooked the differences between elite and commoner among the Mojos. In his manifesto, Daniel Suárez admitted arresting 140 Mojos, but did not mention whipping or killing any of them, or even that they were taken during mass in the church. In the subsequent historiography of the Guayochería, some carayana writers accepted Suárez's contention that the corregidor, the cacique, and Nicanor Cubene had plotted to kill the whites and burn Trinidad.[77]

On June 16, Napoleón Vaca resigned his post of police intendant, and Daniel Suárez appointed Ignacio Aguilera in his place, later asking for the minister of government's approval. The prefect justified the appointment by claiming that the crisis with the Mojos was not yet over.[78] In response to the continuing crisis, an "assembly of notable residents" (*junta de vecinos notables*) met on June 17, whose program was published in Trinidad by the press of Francisco Suárez y Hermanos. In addition to the six members of the Committee of War, this document bore the names of seventy-five other carayanas, including Francisco Valderrama, Nicomedes Bascopé, Nicanor Vaca, and the gringos Samuel Steir and Juan Roberson. Antenor Valverde even signed, having not died in Velasco's expedition after all. These citizens declared that they had met at the invitation of the prefect, with the object of agreeing on measures "conducive to the complete reestablishment of the order altered by the uprising of the indigenous race, [from] whose excesses and tendencies to exterminate the white race the country is suffering." A note dated June 28 added that despite being "invited with insistence," Father Egüez did not attend the meeting. The secretary of the municipal council, Pedro Manuel Silva, also absented himself, while Rosendo Gutiérrez scarcely attended but approved of the junta's resolutions.[79]

The notables agreed on six points, the first of which was to send another expedition to secure the rivers and roads "from the gangs of Indians who set traps for the vecinos and passengers," evidence that Mojo resistance was not completely broken. Second, they were to ask the national government for aid in the form of a military detachment to "pacify and impose respect for the national religion, laws, and flag." Third, they would ask the national government for fifty rifles. Fourth, they would form the junta de vecinos notables, with the prefect at its head, and impose military law for the duration of the crisis. Fifth, they would use the rebels' goods and persons to indemnify those carayanas harmed by the rebellion, and to repay the expenditures of the departmental treasury. And sixth, they would treat the rebel leaders severely, but offer peace to other rebels provided that they appeared at an appointed time and gave up their arms. A final point was added in defense of the prefect's actions, saying that the Indians had prepared their rebellion for three or four years, and that neither the present nor past departmental authorities were responsible for inciting it. Rather, the carayanas traced the rebels' motives "to the inveterate hatred which they have for us and to their savage tendencies manifested at all times since the epoch of their conquest," adding that any other view published in the national press was wrong. In this sweeping justification, then, the carayanas demeaned the Mojos as savages and claimed the right to dominate them through a conquest which never occurred. The prefect's position was also reflected in a phrase of the second point, which said that if troops were not sent, the national government would be responsible for "the fatal events that might arise from its inaction."[80]

## The National Reaction

By the time the junta met, there was little danger of further fighting, despite their call for another expedition. Soon afterward, Daniel Suárez would begin to feel the sting of public criticism of his handling of the events of 1887. Antelo wrote his account of the Guayochería on July 4, which placed most of the blame for the crisis on the prefect. On the same day, Suárez wrote to the minister of government and offered his resignation.[81] But by that time, word of events in Beni had finally reached President Gregorio Pacheco, who had already decided to replace Suárez with a new prefect.

José María Ferrera's letter to Ricardo Chávez was the first report to reach the outside world, and it had arrived in Santa Cruz before April 23. By that date, the prefects of both Santa Cruz and Cochabamba had written to inform the national government about the troubles in Beni. On April 25, the national representative from Beni, Rómulo Arana Peredo, who was in Santa Cruz, wrote to the prefect of that department, Ricardo Arias, asking that he organize a military force to help the carayanas in Trinidad. On April 27, Arias denied Arana's request on the grounds that neither Daniel Suárez nor the national government had asked for help, and on May 14, he wrote to the minister of government, José Manuel del Carpio, to inform him about the representative's proposal.[82]

By April 29, news of the Guayochería had reached the national capital, Sucre. On May 7, Carpio wrote to Arias, asking for information on how best to send a military force to Beni, and "about the conduct of the prefect, intendant, and other authorities of Beni, which according to data, are the cause of that uprising." By May 15, the Sucre newspaper *Los Debates* claimed that Daniel Suárez had asked the government for a force of a hundred men to help fight the Mojos. On May 21, a gentleman from

Santa Cruz wrote to a friend in Sucre about the events in Beni that he had ascertained after receiving a private letter from Daniel Suárez. *Los Debates* published the letter on June 11, but removed the names of the two gentlemen. Not surprisingly, the letter reflected Suárez's views of the situation: that the fighting was a result of years of oppression of the Indians by patrones, and was not the fault of the Suárez administration; and that Suárez's political enemies had been using the defeat of the first expedition to attack the Beni prefect's conduct.[83]

By the end of May, the national government was ready to act and had decided to send an armed force to Trinidad. President Pacheco spoke before the National Congress, telling the representatives that "the government has chosen new authorities, organizing at the same time a competent expedition commended to honorable persons, and hopes that order will be reestablished soon by gentle and conciliatory means." He also said that the crisis in Beni was the "natural consequence" of the abuse of the Indians as labor during the rubber boom, and had no political character, for which Daniel Suárez later called Pacheco his "political friend."[84] On May 28, Carpio wrote to both Arias and Suárez, informing them that the national government was sending José María Urdininea to be the new prefect of Beni. He would be escorted by twenty-five cavalrymen with eighty extra rifles from Sucre to Santa Cruz, where eighty troops from that city would join them for the march to Trinidad. On June 7, the minister of government officially appointed Urdininea, a lawyer and minor bureaucrat from Sucre, who had served as president of the Mesa Inscriptora (Registry of Voters), as the new prefect of Beni. On June 15, Urdininea left Sucre at the head of a small force of men from the Bolívar Regiment. To provide some spectacle, the Loa Battalion accompanied them as far as the city limits.[85]

On June 30, the national government invited three Jesuit priests who had recently been expelled from Peru to go "to employ persuasive and conciliatory means to subdue the rebels of Beni, before applying violent recourses." These priests, Gumercindo Gómez de Arteche, Gabino Astrain, and Ricardo Manzanedo, accepted the offer and traveled a different route than that taken by Urdininea. They left Cochabamba on July 4, descending the eastern front of the Andes until they reached the Chapare River, which they navigated until reaching the Mamoré on July 23. At the port of Limoquije, they encountered Isaac Michel and his family, who were on their way to Santa Cruz. Arteche assumed that the council president left Trinidad "frightened no doubt by his work, that is, by the atrocities committed with the Indians." From the Mamoré the Jesuits continued on to Loreto.[86]

On August 12 they left Loreto and arrived in Trinidad the next morning, where they were offered lodging by Victoriano Bejarano, whom they called "the richest man in the capital and very fond of the Indians," even though he had been part of the junta de vecinos notables that planned the further suppression of the revolt. Daniel Suárez and Napoleón Vaca visited the fathers in the afternoon, and the prefect told them that as the carayanas had violence done to them, "he had executed or permitted unjust and cruel punishments against the poor Indians." Few other carayanas visited the priests, but a large number of Indians came to see them, bearing presents as simple as earthen jars of water. On August 14 the three Jesuits held a service in the Trinidad church, and several of the most important carayanas did not attend. Afterward, Daniel Suárez raised a pennant and led a procession around the plaza, in which macheteros danced.[87]

Suárez advised the priests not to go in search of the Mojo refugees until Urdininea's arrival, so they decided to visit San Javier and San Pedro, which were without priests. In the latter town two carayanas, "badly compromised with two Indian women," failed to attend church, and "an unhappy mestizo, one of the most guilty in the expedition against the Indians," left for the countryside with his family.[88]

Meanwhile, Urdininea continued his journey, leaving Santa Cruz on August 10, and stopping at the Guarayo missions for eight days to rest his troop. On September 7, he rode into Trinidad with his escort of twenty-five soldiers, but without the eighty additional troops which Santa Cruz was supposed to have provided. Arteche says that the Indians were very happy with the new prefect's arrival, but that not one carayana went out to meet him. "They knew well that his arrival and ours signified a solemn condemnation of their conduct on the part of the government and of Bolivia," the Jesuit commented. Urdininea and his secretary went at once to meet with the three fathers, and held long talks with them on that and the following day, in which it was decided that the priests would cross the Mamoré to minister to the Mojo refugees, and to prepare them for a visit from the prefect. On September 9, Urdininea formally took over the prefecture from Daniel Suárez, and on September 17, he wrote to the minister of government denouncing "the constitutional infractions and the iniquities and cruelties that an entire citizenry with the exception of five or six persons, represented by a 'Committee of War,' exercised with the Indians taken in church on Ascension Day." The new prefect added that "it is important to disarm the whites, among whom there are firms like that of Suárez (Francisco), which is said to have up to 200 rifles and people at its disposal."[89]

Urdininea employed the discourse of civilization and barbarism to condemn the carayanas' actions as a "retrogression of civilized people to barbarism," reversing what local carayanas had said about Mojo "savagery." Though taking the Indians' side, Urdininea characterized them as "incapable of being citizens for still not having the capacity to govern themselves; they are children in their character and need the tutelage and protection of the missionary who serves them as father and defender of their rights." In other words, he repeated the Spanish colonial discourse on the Indian, using it to justify government action to protect these "children" while denying their equality as citizens under liberalism. In this respect, Urdininea was quite similar to Daniel Suárez, who claimed that liberal laws freeing Indians from government tutelage had only brought about the labor exploitation of "these free men without intelligence or conscience." To him, Beni Indians were undeserving of the liberty the government had granted them, for it made them more inclined to laziness, drunkenness, and the swindling of patrones' cash advances. Suárez never referred to the Mojos as citizens, instead calling them Indians, indigenes, the indigenous race, or the aboriginal race. The carayanas, in contrast, were citizens, vecinos, whites, or the white race.[90] In the discourses of both the incoming and outgoing prefects, it was now inconceivable that Indians could be citizens.

## The Jesuits' Journey West of the Mamoré

On September 10, the three Jesuits left Trinidad, receiving tearful farewells from the few Indians in town. They reached the west bank of the Mamoré by canoe, and slept at one of the few estancias that had not been burned down. The next day they arrived at Trinidacito, which with another place called Santa Rosa was the only settlement left standing to the west of the Mamoré. The Indians at Trinidacito fell on their knees crying, "without a doubt remembering the great evils they had just suffered, and seeing in us the image of their old protectors," as Arteche wrote. There were few people in Trinidacito, and those had recently come out of their hiding places on hearing that the Jesuits had crossed the Mamoré. They told the Jesuits that the bulk of the people were six days' journey away, deep in the woods. More than five hundred Indians eventually arrived in Trinidacito to receive communion from the fathers. There, the three Jesuits decided to divide their efforts. Astrain would go to San Ignacio to minister to the people there and to try to recover the goods and people taken in Saavedra's expedition. Manzanedo would remain at Trinidacito with their portable altar, and Arteche would go in search of the refugee Mojos in the woods.[91]

In the morning of September 12, Arteche left with two guides and the sacristan of Trinidad, all mounted. Their day's travel was difficult, as they had to pass through several flooded areas, in one of which Arteche's mule sank to its ears. They found only twenty people on the trail, all of whom were heading for the woods when they were detained by the floods. At night, Arteche and his companions arrived at San Francisco, where they found all the houses and the chapel burnt, and had to sleep on the ground.[92]

At daybreak of September 13, the travelers left San Francisco. At 9:00 a.m. they encountered a group of a hundred people taking foodstuffs back to the woods in eight carts, who had been detained at a river crossing. They all surrounded Arteche, who tried to persuade them to join the two other fathers in Trinidacito, saying that they would be safe and that the government, the new prefect, and the sympathies of Bolivia were on their side. Arteche furthermore urged them to return, "chiefly since the cross and fields that they were looking for were a false thing, and to my mode of understanding, a false tradition." This search for a cross and fields is strikingly similar to the goal of the later Loma Santa millenarian movement which manifested itself in several Mojo migrations from 1932 into the 1990s, and because of this reference Guayocho is sometimes taken to be the first Loma Santa messiah. Perhaps the legend already existed among the Mojos of this enchanted isla which could be found by exploring to the southwest, where the Indians would find a cross, be received by Saint Michael and a priest, and live in an earthly paradise with an abundance of cattle and no carayanas.[93]

After hearing Arteche's pleas, one man then spoke to him in Mojo, and the speech was translated into Spanish with the following meaning:

> Many are the vexations we have suffered from the carayanas; they take us upriver and downriver, and to the Madeira itself where so many perish, for whatever they wish to give us, and if we don't serve them to their taste, they give us the lash; now they haven't even left us a palm's breadth of land in which to make a chaco or raise a house; from them we only learn vices, above all drunkenness; they ridicule and impede our Christian feasts and practices with force, and in this manner the people will quickly lose their religion entirely. For whatever your and the prefect's coming does, he and you will go, and we will be left at the mercy of our enemies again. Thus we have resolved to abandon our

lands to the carayanas, the lands of our fathers, and go in search of new ones, that God may give us to understand where.

Arteche realized that he could not convince the Mojos to return to Trinidacito, and after talking with them for an hour and a half, he and his companions continued on their journey. The travelers slept on the ground among the burned houses of San Lorenzo, accompanied by some five hundred cattle that had gathered there of their own accord.[94]

The next day, Arteche and his companions went beyond San Lorenzo in search of more refugees. At one point, they caught sight of two Indian women from afar. The women, thinking Arteche's party to be another carayana expedition to capture people, dropped their belongings and ran to hide in the tall grass. Arteche's Mojo companions called out to the women in Mojo, and the two, who turned out to be mother and daughter, came out of hiding. When they heard that the whippings had ceased and that the Mojo prisoners would be set free, they set off for San Ignacio to try to recover a boy that Saavedra had taken from them. Arteche told the women that they would find Father Astrain there, and that they could count on him. Arteche's party met no one else until they encountered some fifty people beneath the shade of a group of palm trees, who were busy killing cattle and making jerky to send to people further in the interior. The Jesuit gave out religious medallions to the women among them, and left a print of the Virgin on a palm trunk to be their protector.[95]

At this camp, Arteche met Alejandro Nosa, the brother of Nicolasa Nosa, who had traveled three days from the interior to reach the palm grove. From what Alejandro told the Jesuit, it is apparent that his people had fled as far as the territory of the Chimanes, the independent indigenous people who lived to the west of the Mojos. A Chimane named Nicolás had offered Alejandro the produce of some of his own chacos.[96] The next morning, Arteche went to accompany Alejandro to his refuge, taking only his mule and that of the sacristan. In a half hour they had to cross a flooded area, after which they came to a place strewn with pieces of carts and trunks, where Saavedra's men had caught up with some fleeing Mojos. They next reached a dangerous flooded area, in which three oxen were stuck with only their horns and backs visible. Unable to cross, Arteche and Alejandro returned to the palm grove and set out again on foot. Taking a path which appeared like a tunnel in the forest, in a quarter hour they reached some huts on the Río Tijamuchi. There they encountered twenty people caring for a sick man, and three others arriving from further in the interior. Following the river, they continued until dark and slept where night found them.[97]

Before noon on September 16, Arteche and Alejandro crossed the Tijamuchi. Alejandro found a small mule ("like a goat," according to Arteche), which he presented to the father as a mount. Arteche thanked the Mojo, for his feet had become blistered from walking. Nor was this the only discomfort, for the Jesuit complained: "I went covered with forest ticks, up to taking them from my very eyelashes." The travelers next arrived at a new chaco, cared for by a few women with their children. They gave Arteche four eggs, which he realized must have been a lot to give considering the scarcity of food.[98]

Continuing on during the next day, Arteche and Alejandro came across "a few men who came urging on an enormous ox harnessed and loaded with meat" in the afternoon. The men gave Arteche some meat, and headed toward their settlement with him and Alejandro, sending a youth ahead to announce their arrival. They came to a high, flat place where a group of old people met them. The aged Mojos fell to their knees and cried when they saw Arteche, and kissed his hand while he offered some comforting words. Then everyone continued on in single file, led by the old women who were crying and singing, and who were followed by the old men who walked silently and with their heads uncovered. They passed other Mojos on the path who were crying and on their knees, who afterwards joined the file. Finally they arrived at a group of huts beside a peaceful river. This settlement was probably the one named San José in an official report of the following year, founded at the site of the abandoned 1840s settlement, San José de la Sierra.[99]

Arteche stayed with these refugee Mojos for five days and was lodged next to the chapel. He remarked that "the first thing they always strive for is to have their chapel," which must have been true for such a new settlement to already have one. The father found the chapel to be small but clean, with an altar and images of saints and the Christ Child. So even in these difficult circumstances, the Mojos took special care to preserve the material culture that embodied their spirit protectors. Arteche guessed that there were six hundred people there or further in the forest, and that another four hundred must have been headed there. The settlement was full of old people, women, and children. Arteche estimated that there were two hundred children there, adding figuratively that they had been "hidden in the lairs of jaguars to free them from the clutches of men!" He found the Mojos looking sickly, but nevertheless happy in their freedom, living as families and never quarreling. The Jesuit performed a great many religious duties in the five days, including officiating at thirty weddings and about as many baptisms. He was impressed that all of the children he baptized were legitimate, and that scarcely any of the couples that he married were cohabiting before marriage. He wrote to Father Manzanedo in Trinidacito, asking him to bring the portable altar to the ford of the Río Tijamuchi.[100]

On September 23, Arteche left the Mojo settlement with everyone except the very old and the women with small babies. He tried to bring his little mule, but the animal lay down and would not get up. The priest had to continue on foot as before, stopping to give confession to two groups of Mojos he encountered, and to baptize four of their children. At the ford of the Río Tijamuchi, some of the Mojos remained behind, while others continued on with Arteche. At the edge of the forest, the Jesuit met a party of more than a hundred Mojo families coming from Trinidacito. They told Arteche that Manzanedo had stayed behind in Trinidacito, but that some young men were bringing the portable altar as well as sugar, coffee, and rice for him. Arteche noted that when the Mojos stopped to rest while traveling with their images, they would take them out of their trunks and set them up on some sort of table as best they could, even lighting candles before them. Thus when Arteche found his altar, it had been placed on a table beneath an awning, both of which were well adorned.[101]

Manzanedo in fact had sent the altar and food before receiving Arteche's letter, but after it reached him, he set out and met Arteche before dawn on the morning of September 26. The two fathers performed many religious services for the Mojos that were with them and for those from the palm grove, but soon decided to divide their efforts again. Manzanedo would wait for the rest of the Mojos to arrive from the river, and then go to wait for the other fathers at a port on the Mamoré, either El Masi or Limoquije. He stayed there for four days, giving comfort and performing religious services, and witnessing the fiestas and religious practices of the Mojos. Among these were a three-hour ceremony of giving alms for a dead person and the planting of a cross in the name of Saint Michael. The Mojos stayed awake all night praying, singing, and dancing the machetero dance before the cross. They did not want Manzanedo to leave, even after he climbed a termite mound to bless the whole group, so that the priest finally had to simply turn his back and spur his horse on, despite the Mojos' lamentations. In Trinidacito, El Masi, and Limoquije, Manzanedo continued to give communion, confession, and baptism to many more Indians.[102]

The seemingly miraculous arrival of kind Jesuit priests among the Mojos so soon after the brutal actions of the Suárez government, and the setting up of a cross to Saint Michael, can both be evaluated in terms of Loma Santa beliefs. If the Mojos already expected a priest to be at their terrestrial paradise, this explains their reluctance to let Manzanedo leave. More likely, the Mojos only incorporated the idea of a priest at the Loma Santa because of the brief visit of the three Jesuits in 1887. In that case, Saint Michael only entered Loma Santa belief because Manzanedo happened to be among the Mojo refugees during ceremonies for Saint Michael's Day (September 29). The Loma Santa ideology thus owes much more to memories of 1887 than to a supposed continuity from precontact beliefs through the Guayochería that some authors have claimed. The traumatic year of 1887 found the Mojos seeking a safe haven out of the carayanas' reach and finding, and then losing, the care of Jesuit priests. The Mojos' desire to recover the refuge and priest that had slipped through their hands in 1887 certainly contributed to later Loma Santa beliefs. They were also clearly an expression of the same desire for pastoral care among Trinitarios west of the Mamoré that had led Guayocho to request a priest for San Lorenzo.

Arteche left the camp at 4:00 p.m. on September 26 with the intendant and eleven other important Mojos (*principales*), who were to meet the new prefect and ask his advice regarding their plan to remove themselves to new lands. The Jesuit writer does not give the intendant's name, but he could have been the erstwhile Santiago Tamo, who had disappeared during the first carayana expedition. The Mojos who stayed behind asked Arteche to appoint someone to govern them while the intendant was away, and when the Jesuit left the choice up to them, they chose Prudencio Nosa, the father of the aforementioned Nicolasa and Alejandro Nosa. Prudencio Nosa was fifty-six years old in 1887 and had long experience on the west bank of the Mamoré. In 1861, he lived at the estancia of Rosario with his wife Ursula and their children Bernabé, Nicolasa, Angel, and Alejandro. The Nosas must have known Andrés Guayocho and his family, who were agregados at Rosario in that year. By 1867, however, Prudencio Nosa was living in Trinidad as a member of the manzana of Jenner, and it is unknown when, or how many times, he returned to the west bank. His mature age, together with his connections to the land and people of the west bank, made him a natural choice as leader of the refugees. If he had maintained some sort of relationship with Guayocho, the refugees had another reason to respect Nosa.[103]

Arteche headed to San Ignacio first, with the intention of redeeming several married women held captive by Montenegro, and some adolescents and children taken by Saavedra. The trip to San Ignacio was eventful, as they met two more caravans of Mojos headed for the backcountry, saw plenty of deer and rheas, and crossed several rivers, once in an oxhide boat. Arteche wrote of the rivers that "even though caimans, otters, and dolphins abound in them, none of these hurt us, this being reserved for a daring piranha, a small fish which, without further ado, took a good bite out of the arm of one of the commission."[104]

Arriving in San Ignacio, Arteche performed a mass in the old Jesuit church and then sent a "tender letter" to the estancias of Montenegro and Saavedra, hoping to convince them to release their prisoners. Montenegro gave up two women and a boy, but Saavedra "contented himself with a courteous reply full of nice words." Before this, Father Astrain had also gotten nowhere with Saavedra: after finally managing to meet the expeditionary, the priest had preached to him for two hours and only obtained Saavedra's promise that he would go to Trinidad to set everything straight. Arteche and his companions left San Ignacio that same night, accompanied by music and lights from the people of the town. Traveling all night, they arrived back in Trinidad on October 1 before noon, where Father Astrain was waiting.[105]

The three fathers left Trinidad for Cochabamba on October 3. The Trinitario refugees west of the Mamoré had been living in fear when the fathers first arrived. The refugees hid from the carayanas, tried to form new settlements, and sent out parties of men to kill cattle for meat to keep them from all starving, while women, children, and old people remained hidden. The Jesuits and Urdininea had taken the first steps toward convincing the Mojos that they could come out of hiding and return to a more settled existence. What became of the Trinitarios who had survived the brutal suppression of the Guayochería to the west of the Mamoré will form the subject of the following chapter.

## Conclusion

The events of 1886 and 1887 were important manifestations of overt Mojo resistance to their exploitation during the rubber boom. Migration was their preferred form of agency and was a viable option because of the llanos environment and the frontier quality of Beni. They were willing to abandon Trinidad in order to escape the carayana government, which served the rubber interests, and their own cabildo, which served the carayana government. The appearance of the Itonama shaman Andrés Guayocho among them added a new dimension to their agency, as he articulated a millenarian ideology that gave supernatural sanction to the abandonment of Trinidad and the creation of a new cabildo in San Lorenzo. Though

Prefect Daniel Suárez justified his repressive actions as a defense of the Bolivian nation and its official church, his policies are better explained as an attempt to recapture labor for rubber extraction and river transport. The prefect succeeded in killing Guayocho and in disrupting whatever organization his ideology brought to the Mojo refugees. But because the national government disavowed Suárez's policies and decided to act (in colonial fashion) as protector of the Indians, he did not bring an end to Mojo agency, or even to the liberty of a large group of those Mojos who had left Trinidad behind.

CHAPTER FIVE

# *The Citizen Cacique*

## After the Guayochería

In September 1887, Andrés Guayocho's community on the isla of San Lorenzo lay in ruins, its burned shells of buildings and silent plaza returning to cattle pasture, its people dead or dispersed. Few could have guessed that within a few years, that same isla would host a revived Mojo settlement that would not only achieve Guayocho's goal of autonomy from the carayana and indigenous authorities of Trinidad, but maintain it for three decades. The new community would not appear overnight, however, and for information on the early movements of Mojo refugees, we must turn to carayana sources from Trinidad.

Most of the principal carayana actors in the Guayochería disappeared from the scene quite rapidly. On October 25, Napoleón Vaca died of a "malignant fever," and on November 5, Daniel Suárez died from the same illness.[1] Miguel Antelo may have remained in Trinidad into the early 1890s, but he eventually moved to Baures, where he became an important patron by 1902, and corregidor by 1906.[2] José María Urdininea did not remain in Trinidad for long either. On January 1, 1888, he wrote to the minister of government saying that it was necessary for him to resign the post of prefect owing to a dislocation he had suffered, and that he would appoint the police intendant Marcelino Marañón as interim prefect. Urdininea stayed in office until the end of March, when he received word from the minister that he had been named prefect of Santa Cruz. He eventually returned to Sucre, where he became a respected writer and public figure, and where he died in 1909.[3]

Despite the end of the fighting, many Trinitario Mojos continued living far to the southwest of Trinidad. On January 23, Urdininea wrote to the minister of government about the peaceful situation in Beni, and noted: "The natives remain submissive and very deferential to authority, devoted to their farming and manning boats, without noting the smallest indication of revolt among them, even though a part are still far away on the banks of the Río Sécure after the flood of 1853, for other motives, and lately because of the turbulence and hostilities of last year, causing a scarcity of hands for work, especially in this capital."[4]

The situation had begun to change by July 9, when acting prefect Marañón informed the minister that some of the Mojos were moving closer to Trinidad once again. Marañón wrote that in February, some fifty emaciated Indians had left the refugee settlement of San José because of diseases and the sterility of the land, and had settled in San Ignacio. In June, Indians from San José moved to San Francisco, Rosario, and Recreo on the west bank of the Mamoré, and sent emissaries to Trinidad "to convince themselves that they are guaranteed their constitutional rights, in order to be able to accommodate themselves in the places of farming and pasture that they possessed." Prudencio Nosa, who had been elected leader of one group of refugees in Father Arteche's presence, was still respected among them and went to speak to Marañón in Trinidad. The acting prefect advised Nosa to return to enjoy the fertility of his properties, to exchange his produce for "articles necessary for social life," and to live closer to the Mamoré and Ibare rivers, with their commerce with the outside world. Nosa said that his people wanted exactly that and added that he wanted to reclaim his grandchildren, the children of his daughter Nicolasa and Nicanor Cubene, "who remained orphans and they and their possessions in the power of strangers."[5]

At the end of 1888, the new prefect, Santos M. Justiniano, informed the minister of government about the situation of the Trinitario refugees. The prefect wrote that he had told the Trinitarios "that in the future they will enjoy constitutional guarantees, and their children will have easy access to primary and secondary education."[6] In other words, he promised them that they would be able to enjoy the full rights of citizens under liberalism, which is exactly what they wanted. Many refugees, including some of the Nosa family, accepted the offer. Although the fate of the Cubene children is unknown, both Prudencio Nosa and his son Alejandro were satisfied that it was safe to return to the west bank of the Mamoré. In 1902, Prudencio owned an estancia named Mercedes with 50 cattle, and Alejandro had a property named Ychimbu with 25 cattle.[7] By 1912, Alejandro Nosa was proprietor of Mercedes, so it is likely that his father had died.[8] Alejandro still held property in the area in 1925, when he owned the estancia of

Santa Cruz and 200 cattle. However, his elder brother Bernabé stayed a greater distance from the authorities, keeping 211 cattle at the estancia of Concepción near San Lorenzo in 1925.[9]

In December 1888, Justiniano wrote that the majority of Mojos had returned to their homes, "with the exception of those great criminals who until today remain rebels."[10] These "rebels" were other groups of Trinitarios who remained in their refuges and only returned to the west bank of the Mamoré in the 1890s. A recent secondary source called the *Historia del pueblo Mojo* contains an account of a group of refugees who returned to resettle San Francisco, which is drawn from oral history. According to this source, a group of Trinitario boatmen fled toward the Río Sécure with their religious images after the events of Ascension Day in Trinidad in 1887. They settled at an isla near the Río Chiripopo and named their settlement San Bernardo. One of them, Pascual Conotiri, bore the title of *maestro de capilla* (master of the chapel) and ordered that a structure be built to house their images. One day, three men from San Bernardo went hunting and were captured by a carayana expedition. The three, Manuel Moy, Pedro Muyba, and another whose last name was Masa, were taken to Trinidad and tortured for many days so that they would reveal the whereabouts of other refugee Mojos. Eventually they escaped and fled with some of their relatives to the Río Chiripopo and found it flooded, with little dry land for so many people. Despite this initial setback, this group of Mojos continued to live as refugees for several years under the leadership of Moy and Conotiri.[11]

In 1897, for reasons that are unclear, these Mojos reoccupied the site of San Francisco. There, they created a cabildo, an orchestra, and a choir, and celebrated their fiestas. The Indians cultivated chacos individually, but they tended cattle, spun thread, and wove cloth communally. They sold their produce in Trinidad and bought other goods there, which were shared among the community.[12] A revival of communal work and ownership of cattle could have been an attempt to reject liberal reforms and return to the way things had been done in a mission town. Such communalism is called into doubt, however, by the fact that a 1925 *catastro* (register of rural properties) lists both Moy and Conotiri as owning their own cattle.[13] San Francisco community members certainly registered their cattle as individual property to comply with Bolivian law and benefit from the protection of property rights that it afforded them. The present evidence does not allow us to determine whether they managed their cattle communally in actual practice, or whether communal control is a latter-day idealization in the oral tradition.

However production and distribution were managed, San Francisco had been reestablished, though the *Historia del pueblo Mojo* states that flooding required its translation to a property owned by Mercedes Noco de Nosa in 1931. San Francisco was laid out like a classic mission town; the same source contains a recent hand-drawn plan of the town as it was in the early twentieth century, with a plaza and its central cross, and a church, parish house, school, and several houses around its perimeter.[14] A Franciscan primary source from 1919 corroborates some of the account given in the *Historia del pueblo Mojo*. Before 1919, according to the Franciscan report, San Francisco was moved from an unhealthy and damp location to a new isla two kilometers further north. There the Mojos built "a solid chapel and houses around a plaza."[15]

## José Santos Noco Guaji and San Lorenzo

Another group of refugees came to reoccupy the isla of San Lorenzo. Its leader was a Trinitario named José Santos Noco Guaji, who is usually referred to simply as Santos Noco. The matrícula of 1874 lists him as an originario of Trinidad, twenty-five years old, married to Tomaza Semo, and a member of the manzana of Ayacucho.[16] How he came to lead a group of refugees is uncertain. The *Historia del pueblo Mojo* notes that some people said Santos Noco was a follower of Guayocho, while others believed he had been a soldier in Saavedra's expedition against the shaman.[17] The confused name "Andrés Noco," which appeared in the Sucre press in 1887, might be evidence that Santos Noco was a close associate of Andrés Guayocho. Since the second manzana, Ayacucho, was the same as the second parcialidad, Sacristanes, then Santos Noco could have had some connection to the west bank of the Mamoré through the Sacristanes' estancia of Asunta. But since Santos Noco appears in none of the contemporary accounts of the Guayochería, none of these assumptions can be proven. Most likely, Santos Noco was neither the choquigüero's ally nor his enemy, but (like Pascual Conotiri) simply fled Trinidad after the events of Ascension Day.

Exactly when the refugees led by Santos Noco returned to San Lorenzo is also unclear. Oral histories collected in San Lorenzo in the 1990s give the year it was "founded" (actually resettled) by Santos Noco as 1887 or 1888, a date which is probably too early.[18] The year 1892, provided by one secondary source, is a more likely date for the resettlement, though the primary source of this information is not cited.[19] The first mention of the resettlement of San Lorenzo comes from the Italian Franciscans Sebastian Pifferi and Zacarias Ducci, who visited all of the missions of Bolivia in 1893. Passing through Trinidad, they obtained the following information:

> A great part of the Indians have gone to inhabit the other bank of the Mamoré, at a place named San Lorenzo; and when they are invited to return to their town, they respond: Why go looking for death? They are very religious, and tenaciously conserve the practices and customs established by the Jesuits. They have shown much desire to have a priest to assist them in spiritual matters; but in their new town, where they have built a chapel, in which they assemble on Sundays and feast days, they pray the Christian doctrine and sing the divine hymns, as they did in Trinidad.[20]

Although Pifferi and Ducci made no reference to Santos Noco's presence in San Lorenzo, the carayanas of Trinidad knew about him by 1894, when they regarded him as leader of the refugee Mojos and as Guayocho's successor.[21]

Also in 1894, the Bolivian government incorporated San Lorenzo into its administrative structure, at least nominally. A national law of October 22 of that year named San Lorenzo a vice cantón of the Province of Cercado (which comprised the Mojo and Canichana towns), and its subcorregidor was presumably Santos Noco.[22] The Mojo leader undoubtedly accepted this situation because he gained an official title and recognition of his people as citizens without having to give up control. The government recognized San Lorenzo as a vice cantón to try to win Santos Noco's favor, and because the only way to "incorporate" the settlement into the state was by assigning it a rank in the administrative hierarchy. However, giving San Lorenzo such a name did not bring it any more under state control than it had been before.

For much of the 1890s, Santos Noco quietly rebuilt San Lorenzo. In 1899, he learned that the Franciscan missionary Fray Daniel Carvallo had brought fifty neophytes from the Guarayo missions to Trinidad to work on certain government buildings at the request of Lucio Pérez Velasco, the vice president of the municipal council. One day in October, when Carvallo was about to leave Trinidad and return home with the Guarayos, the prefect received a letter from Santos Noco. "I send a canoe with four men," the Mojo leader wrote, "so that the *tata fraile* [father friar] may come; my cabildo is glad to have him with them, and for this I send [it] soon; the canoe, though small, is solid; with the father friar may come you and Mr. Francisco Vargas and no one else." Carvallo was not able to go to San Lorenzo, but he wrote to Santos Noco promising to meet his people at a later date.[23]

This episode provides us with the important information that San Lorenzo had its own cabildo again by 1899. The cabildos in Santos Noco's San Lorenzo and Moy and Conotiri's San Francisco were direct descendants of Guayocho's cabildo in San Lorenzo. The later leaders, just like Guayocho, wished to declare their autonomy from the Trinidad cabildo, which had recently served the carayanas as an instrument to exploit Mojo commoners. In addition to freeing the refugee Mojos from such direct exploitation, the creation of new cabildos allowed them to take control of the all-important religious fiestas, services, and dances away from Trinidad. By constructing churches and plazas at the refugee towns, these Mojos provided a more or less scattered population of slash-and-burn farmers and cattle herders with local ceremonial centers. This in turn made it unnecessary for them to travel to Trinidad and expose themselves to the demands of the authorities. From the 1890s until the present day, Mojos have continued to found cabildos at new settlements for much the same reasons. Sacristanes, abadesas, and macheteros all survived in the new settlements, and were well able to handle most religious matters on their own. Even so, the refugee Mojos were devout Catholics and always wanted ordained priests to perform marriages, baptisms, and confirmations. Just as Guayocho had asked for a priest for San Lorenzo, Santos Noco tried to get Father Carvallo to go there and would continue in his efforts to secure priestly care of his people.

Despite Santos Noco's desire for religious contact, he was quite unwilling to submit his people to secular authority. By 1900, San Lorenzo had achieved a certain renown among carayanas as a virtually independent Mojo settlement, "the famous town of Santos Noco," who was its "lifetime Governor." The published results of the 1900 census of Bolivia noted that exact figures could not be obtained for the "extensive region which is under the dominion of the singular Governor Santos Noco," adding that this was "as much for the distance as because no one wants to travel there." Even so, the secretary of the Central Commission for the Census in Beni was able to estimate, through "references of the caciques of this town," that there were not less than five hundred people in the area ruled by Santos Noco. One settlement, Todos Santos, was said to contain fifty schoolchildren and sixty women. The census also continued to refer to San Lorenzo as a vice cantón of the Province of Cercado.[24]

Even though San Lorenzo was difficult to reach, it was not totally cut off from the outside world in 1900. By that year, Santos Noco was writing letters to both the prefect and the priest (*vicario foráneo*) in Trinidad, and receiving news from there in turn, according to the Italian Franciscan Fray Bernardino J. Pesciotti, prefect of missions of the Colegio of Tarata near Cochabamba. In 1900, Pesciotti learned about the Mojos of Santos Noco while traveling from the Guarayo missions to Cochabamba, and wrote:

> On the Río Sécure live many Trinitario Indians, subject to a Cacique Santos Noco, who is the *fac totum* for them, among whom he enjoys a growing power. . . . They are those Indians that escaped from the persecution of the whites in the year 1887. Nobody bothers them. Santos Noco has them at his disposal, lending some crews to persons well known to him, greatly trusted by him, and with the assurance that they will not share the same fate as other Indians on the lower Beni. On my journey I had occasion to become acquainted with several, among them his brother-in-law; two were pilot and *puntero* [another crewman] of my boat: I was able to persuade myself that they have conserved, better than the others, the religious instructions that they received from the Jesuits. I manifested to them the desire I had to visit them, to become acquainted with their Chief, and to live among them or to provide them with a missionary. My proposal was received with admiration, with respect, and with jubilation.[25]

Interestingly, Santos Noco was not averse to lending his people out to work for carayanas under his own terms. If he was a member of the parcialidad of Sacristanes, he may have assumed the right to order his people to work for carayanas and to mediate between the two groups. And if he did assure his people's safety, they would have accepted him as a representative of traditional authority. Who were the carayanas that Santos Noco knew and trusted with his people? One might have been the Francisco Vargas whom he asked to come to San Lorenzo with Father Carvallo. Another was surely municipal council vice president Lucio Pérez Velasco, who, according to Pesciotti, "knew how to win the benevolence of the chief of the Trinitarios" and of his people by giving them statues of Nuestra Señora del Carmen, San Lorenzo, and San José, as well as two church bells and cans of liquor.[26] Much like the hacendados who constructed oratories for the Indians to gain access to their labor before the Guayochería, Pérez Velasco also showed an awareness of the importance of religion among the Trinitarios. It is unclear whether he provided alcohol to the

Mojo workers through Santos Noco and with his approval, or behind the Mojo leader's back. As drinking had long been part of Mojo religious fiestas, it is unlikely that Santos Noco or his cabildo would have objected to the use of alcohol in that context.

Despite the isolation of San Lorenzo, Santos Noco tried to keep abreast of events in Trinidad and the rest of Bolivia. By 1905, he was receiving the Trinidad newspaper *La Democracia* through another Mojo named Visitación Noe. In that year, Santos Noco introduced himself and his views to the outside world by writing to the newspaper's editor. "From this town set apart that my predecessor *Guayocho* founded, searching for a secure asylum for our disinherited race, I direct my first correspondence to your popular weekly," the Mojo leader wrote on July 3, after reading the July 2 edition of *La Democracia*. In that issue, the editor had published a poem by Dr. M. Pacífico Roca, entitled "The Destiny of the Indian Race," to which Santos Noco felt compelled to respond. In thirteen stanzas, Roca wrote how the Indians were destined for extinction after coming into contact with whites.[27] Like other white authors influenced by romanticism, he expressed sympathy for the Indians only when whites felt assured that their culture had been destroyed.

In a response intended to point out that the Mojos and their culture were still very much alive, Santos Noco addressed himself "to the men directing the destinies of the country and to the representatives of the department." The Mojo leader took up the discourses of liberalism and of civilization and barbarism and used both to demand better treatment for his people. He wrote that they belonged to the "Bolivian family" just as much as the Aymaras of the Altiplano, and that they needed several things in order to progress. In the first part of his letter, Santos Noco phrased his petition in the language of liberalism, advocating national unity, education, and technological progress:

> We need schools that will teach *Castilian* to us and our children, because language is a strong bond of union. We need Salesian schools, which will teach us to work according to the latest advances of the age. We need foreign immigration, not to despoil us of our homes, but rather to teach us to exploit the privileged riches of our soil to greater advantage with machines, whose benefits would redound to the profit of all but not to fortunate *syndicates*. That with railroads, telegraphs, roads, and steamboats, we be made participants in the *two million pounds sterling* of Acre [the sum paid by Brazil after annexing part of the Bolivian Amazon in 1903], an integral part of our own soil.

In contrast to the Mojos of the 1840s, who were merely the subjects of liberal carayanas' attempts to reform education, Santos Noco now requested specific types of schools to help the Mojos advance in the modern world. He did not see knowledge of Spanish as diminishing native culture in any way, but rather as a useful tool that would allow the Mojos to participate in liberal Bolivia as equals and to escape discrimination against monolingual speakers of native languages. Nevertheless, Santos Noco concluded his demand that his people should benefit from the progress of the liberal era with a determination to preserve some elements of colonial mission culture. The Mojo leader wrote:

> And finally, that the Catholic religion, which removed us from barbarism and incorporated us into the regime of civilization two centuries ago, in which we remain faithful, conserving our songs and dances, which our true fathers the Jesuits taught us, be promoted among us.
>
> We ask that if the wild cattle have to be divided among the godchildren of the government, that they be adjudicated to us in lots to the distinct *aillos* [Indian communities], as first community members and occupants of the soil.

Santos Noco not only insisted that his people were civilized, but also defined civilization as the form of Catholicism that the Mojos had created in partnership with the Jesuits. He also intimated that it was the Mojos' own peaceful acceptance of Catholicism that raised them from savagery to civilization, not any armed conquest by whites. Santos Noco finished his letter to *La Democracia* by thanking Roca, "beseeching him . . . to continue playing the melancholy notes of his lyre, in order to awaken the dormant sentiments of the *carayana* in favor of our race."[28]

It was not long after Santos Noco established friendly contact with Trinidad that carayanas in the Beni capital began to dream of tapping the labor pool represented by the people of San Lorenzo. In 1906, this idea occurred to the author of an article in *La Democracia*, who observed that the prefect, Dr. Luis F. Jémio, had begun construction on "the missing part of the houses of government." With all of the necessary material gathered, the prefect's only problem was a lack of laborers. The writer suggested that "people from those of Santos Noco" be brought and paid a good daily wage. It seems unlikely that Santos Noco provided such labor, if such a request was ever made, for Prefect Carmelo López had to recruit twenty-five Ignacianos for the construction work in 1907.[29] At any rate, Santos Noco continued to correspond with Trinidad. In the rainy season of early 1907, he wrote to *La Democracia* with the information that the Río Sécure had flooded, and that "the rebel tribe of which he is chief" (in the newspaper's words) had to go in search of higher ground to make new chacos, which happily they had found.[30]

The Swedish anthropologist Erland Nordenskiöld heard more about Santos Noco's San Lorenzo when he visited Trinidad in 1909. He wrote that Santos Noco "now rules over a little independent country with a church and school." According to Nordenskiöld, the Mojo leader permitted white merchants to visit San Lorenzo, but only allowed them to remain for twenty-four hours. Nordenskiöld saw letters that Santos Noco had sent to Trinidad, which appeared to be "excellently conceived." In his letters, the "little Indian king" treated the prefect as an equal, and in one he asked casually about President José Manuel Pando, "What is this Pando doing now? Greet him for me." The anthropologist explained that the Bolivian authorities tolerated Santos Noco's independent behavior because they knew that he and his people would just disappear into the wilderness again if they ever sent an armed expedition against him.[31]

## San Lorenzo and the Franciscans

In 1911, Santos Noco took advantage of the pastoral visit of the bishop of Santa Cruz, José Belisario Santistevan, to Beni to communicate with some of the most important Catholic clergymen in the Bolivian Oriente. After visiting Loreto and Trinidad, the bishop and his party set out on the road to San Ignacio, where they arrived on the morning of June 28. In the afternoon of that same day, Santos Noco appeared in San Ignacio in perhaps his first visit to a carayana-ruled town since 1887. The Mojo leader met publicly with the bishop, who promised him protection from the whites and tried to convince him "to resume dealings, friendship, and commerce with the whites, leaving his hiding place with his people, and to not deprive himself of the benefits of religion, living far from the towns and the priest." Santos Noco appeared to agree, but the next morning he was nowhere to be found. He had come to meet the bishop in San Ignacio rather than in Trinidad, in order to avoid the prefect and the large white population of the capital. Apparently he felt uncomfortable even in San Ignacio, for it was said he left in flight.[32]

Nevertheless, on June 30 Santos Noco sent a letter to the bishop inviting him to visit San Lorenzo. Santistevan did not think it prudent to go there himself, but he offered the task to his companion, the Italian Franciscan

Fray Francisco Pierini, prefect of missions of the Colegio of Tarata, who enthusiastically offered to visit San Lorenzo. Despite some initial misgivings, the bishop agreed to let Pierini go. The Franciscan was well received by the "tribe of Guayochos," as he called the people of San Lorenzo. "Their chief, a certain Santos Noco, who is not wholly a common person and who even possesses notions of writing and reading, assembled the Captains, and together with them, begged me to found a mission among them," Pierini wrote. He had to respond that the Colegio of Tarata was short of personnel, but that he would do what he could to remove any obstacles toward the foundation of the mission. It seems likely that Santos Noco and Pierini then agreed on a strategy to gain government approval for the mission. Pierini would present the "Guayochos" essentially as a new tribe of bárbaros, who would thus be eligible to receive the protection of a mission. After confirming many children and adults, and performing seventy-four weddings and 459 baptisms (some of adults), he returned to San Ignacio, and rejoined the bishop in Trinidad on July 9.[33] The adult baptisms show just how long San Lorenzo had gone without the attention of a priest.

Later, from his base at the Guarayo missions, Pierini wrote to Tarata and asked permission to found a mission at San Lorenzo from the Venerable Discretorio, a council of four Franciscans elected every three years by all the brothers of the colegio. The council promised Pierini a Franciscan for the mission, and he then wrote to the minister of war and colonization (responsible for defense and the settlement of the Oriente) in La Paz. The minister responded that both he and the president strongly supported Pierini's plan "to found a new mission, with a tribe of Guayochos," and added that they would provide financial help. After receiving this approval, and already knowing that the bishop supported him, Pierini informed the minister in December 1911 that he had decided to go to San Lorenzo himself after the end of the rainy season, "hoping to be able to overcome the obstacles that could oppose the realization of a plan, which is to return to Christian society, to the Fatherland, and to commerce, a matter of 800 souls, that in 1887, after bloody combats with the whites, had removed themselves from the influences of civilized life."[34] While relating something of the Guayochos' true history, Pierini nevertheless emphasized that whatever their origins, the people had regressed to barbarism and now needed a mission to bring them back into the fold of civilization.

The Trinidad press began to take note of the Franciscan interest in San Lorenzo in 1912. On June 20 of that year, *El Eco del Beni* wrote that Father Pierini was headed for the region of the Río Sécure "to found the mission of San Lorenzo with the subjects of Santos Noco."[35] On July 2, Pierini wrote to Angel Lara, the fiscal (departmental attorney general) of Beni, to complain that certain residents of Trinidad had filed petitions for pastureland at Rosario, San Francisco, and San Lorenzo, which the Indians "vulgarly called Guayochos" had possessed "from time immemorial." Granting such lands to others, Pierini wrote, would not only be a "patent injustice" but also "perfectly illegal." He cited national laws that provided for the protection of Indian lands, and that the corregidor of the district where land petitions were located had to make sure that no Indian communities possessed lands in the same place. If they did, the corregidor had to warn them so that they could protect their rights according to the law. Pierini later wrote to *La Democracia* to confirm the foundation of the mission.[36] It was obvious that in Pierini, Santos Noco had found an ally skilled in the legal protection of Indian lands.

Pierini had reason to want to defend Mojo lands against carayana land petitioners. On October 26, 1905, the national government enacted the *Ley de tierras baldías* (law of barren lands) to stimulate colonization in the Oriente, partly as an attempt to strengthen Bolivia's hold on the areas that remained after Brazil took Acre in 1903. According to article 4 of the law, the state would sell portions of "unoccupied" land of up to twenty thousand hectares each to any Bolivian or foreigner who obeyed the civil law of Bolivia. The buyers would pay ten centavos per hectare of agricultural or pasture land, or one boliviano (ten times as much) per hectare of land that contained rubber trees.[37] Tierras baldías did not exist under the communal property system of the missions, as all the land was considered to be utilized either by agriculture or cattle raising.[38] Though carayanas had been acquiring land since the end of most communal property in 1856, the law of 1905 made it easier than ever for them to obtain large properties. At first there were relatively few land petitions around Trinidad, but when British rubber plantations in Asia began to produce for the market around 1912, the price of rubber finally began to drop after eighty years of growth. As a consequence, more carayanas seem to have turned to cattle ranching around the former mission towns as a more secure investment.[39] The resulting spread of carayana ranching around Trinidad is what made Santos Noco invite the Franciscans to San Lorenzo as a means of protecting its lands against carayana encroachments.[40]

On August 1, 1912, *El Eco del Beni* protested the creation of the mission using typically liberal arguments. The anonymous author said that Trinidad's commerce would suffer, because some Trinitarios were already moving to the mission, and furthermore that there had been no legal authorization (that he knew of) for the mission. The author then protested the need for a mission among the Trinitarios, citing article 1 of the Regulation of Missions, which stated that the object of the missions was "the preparation of the nomadic elements of the country for civilized life." The author pointed out that the Trinitarios were not nomads but rather were landowners who cultivated crops and raised cattle on their land. Moreover, they had a "special preference for voyages," and many had sold their services as boatmen on the Río Chapare or to Santa Cruz (steamboats now dominated river traffic below Trinidad), after asking for cash advances. The Trinitarios were thus already "free men," and already lived a civilized life, which consisted of the exercise of civil and political rights. A mission at San Lorenzo would "upset this order of things" and make the "poor indigenes" regress seventy years. The author concluded with the statement that "since Bolivia, a country bound to the old Spanish usage, can't live without friars, [then] place a father conversor in San Francisco, San Lorenzo, or Candelaria for the Guayochos, who aren't savages, but only refuse to come near us since the historic uprising of '86."[41] On August 15, another article in *El Eco del Beni* offered the liberalism-tinged comment that the Trinitarios "still conserve certain primitive customs and tendencies, which have given a reason for the partisans of medieval civilization to bring us a mission of Franciscan friars to the outskirts of Trinidad."[42]

At the beginning of August 1912, Fiscal Angel Lara went to visit the new mission of San Lorenzo and wrote a description of the town. It had been "founded by the Indian chief Santos Noco on a beautiful plain, regular and high." From the high ground of Santos Noco's isla, one could see the horizon in all directions, and even to the first red granite foothills of the Andes. Near the town were forests suitable for slash-and-burn agriculture, which also furnished timber for construction. The country round about appeared ideal for raising cattle and pigs. The town of San Lorenzo was "well-formed, clean, and methodically constructed." Its spacious plaza was the center of a settlement that included a church, Santos Noco's house, and a school where forty boys learned to read and write Spanish and also something of music as well. Girls did not attend school because they were only taught to pray and to perform domestic chores. Just as Guayocho had kept a Bolivian flag in San Lorenzo, so Lara found that Santos Noco took great care of one Bolivian coat of arms in his house, and of another in the school. The church choir also sang the Bolivian national anthem for the fiscal.[43] Through all of these manifestations of loyalty to the liberal Bolivian state, the Lorenzanos demonstrated their worthiness to enjoy the equal protection it promised to all its citizens.

Lara wrote that "Santos Noco is undoubtedly an intelligent man and a man of character, attentive with his guests. He speaks Spanish and is very fond of reading newspapers." After meeting the new Franciscan stationed at San Lorenzo, an Austrian named Juan Félix Jenewein, the fiscal called him a "young man of very estimable talents and austere customs." Lara noted that Father Jenewein would not accept Trinitarios in his mission who owed labor to carayana patrones for two reasons: firstly to avoid harming the commerce and industry of Trinidad, and secondly to prevent the "very free customs" of those people from demoralizing the Indians of San Lorenzo. According to Lara, the Lorenzanos were very respectful and obedient to Jenewein, and very happy that they now had a priest. The fiscal also observed that the Trinitarios who lived west of the Mamoré were divided over the establishment of the mission. Those of San Lorenzo and its environs (including San Francisco) were happy with the mission regime, but those from Rosario to the Mamoré were against it. Lara wrote that they preferred "the life of Trinidad," and only wanted to work to repay the advances they had received from carayanas, and to cultivate their small properties.[44]

Father Jenewein told the fiscal about his plans for the mission. He planned to expand the agriculture of San Lorenzo from the growing of subsistence crops to include large plantations of sugarcane. As the Lorenzanos had lost about a third of their cattle to a new epidemic called *lengüeta* (anthrax), he proposed to rebuild their herds. And finally, the priest spoke of the establishment of two workshops, one of carpentry and the other of ironworking. He already had the tools and the sites for the workshops selected, and wanted to bring three carpenters and three blacksmiths from the Guarayo missions to teach their skills to the Lorenzanos. Jenewein told Lara that the Franciscans had received two thousand bolivianos from the minister of war and colonization for the founding of the mission, and that he had spent the money in moving himself to San Lorenzo, and in buying food supplies for himself and tools for the workshops.[45]

Lara wrote that "those Indians remember, very moved, the massacre of 1887, and as they say: 'the burning of their images and oratories which were reduced to ashes.'" Even though they claimed to have forgotten their resentment and now confided in the government, the Lorenzanos feared "the nearness of the whites, because these are abusive and absorbent." For this reason, Father Jenewein and Santos Noco asked the government, through Fiscal Lara, for "a zone of land with fields and forests, sufficient for agriculture, cattle raising, and the natural expansion of their populations." They also requested that the Indian lands be well delimited "in order not to be persecuted and annoyed by the whites." Lara estimated the population of San Lorenzo to be two hundred people, or at least five hundred including the nearby settlements of San Francisco and Trinidacito.[46]

On August 30, 1912, President Eliodoro Villazón and Minister of War and Colonization Julio La Faye gave their official authorization for the San Lorenzo mission in a supreme resolution of the national government. They referred to the people of San Lorenzo, Rosario, San Francisco, and Trinidad (i.e., Trinidacito) as Guayochos, and identified them as "tribes that populate the Department of Beni, in the region embraced between the rivers Mamoré and Sécure." The president and minister justified the foundation of the mission as part of the government's duty "to procure the education and instruction of the savage Indians, equipping them, through the colonizing work of the father conversors, for civilized life and the national exercise of the civil rights recognized by the fundamental law of the state." Though seemingly accepting Pierini's Guayochos as a new tribe of bárbaros suitable for civilization in a mission, Villazón and La Faye also identified the people of the towns west of the Mamoré as Trinitarios. Presumably these important government leaders in La Paz were not entirely familiar with the history of the peoples of Beni. At any rate, they gave their approval for the mission, and moreover ordered the prefecture of Beni to survey the boundaries of the Indian lands, and to send the resulting report and map to the Ministry of War and Colonization.[47]

In response to the national government's supreme resolution, *El Eco del Beni* offered a criticism on October 17, writing that "the Ministry of [War and] Colonization has been the victim of a deceit, a deceit that can't be explained except as the fruit of the ignorance of the secretary of state, Mr. La Faye, of the geography and ethnography of Bolivia." The article's author claimed that San Lorenzo, Rosario, San Francisco, and Trinidacito did not need to be colonized, as they were populated by Indians from Trinidad who were proprietors of farming plots and cattle estancias there. Moreover San Lorenzo was the seat of a vice cantón of the Province of Cercado, and thus already incorporated into the state. Continuing, the author remarked ironically that now a liberal government wanted "to reduce the Trinitario Indian, who always exercised the functions of Bolivian citizen, to the condition of NEOPHYTE only to satisfy the wish of the Prefect of the Missions of Tarata!" He suggested that the Franciscans go in search of real savages to convert and civilize, and then accused them of deceiving La Faye, "making him believe that the Trinitario Indians, who bear the nickname of GUAYOCHOS since the year '87, are savage forest dwellers and without a whit of civilization."[48]

The writer went on to say that many of the Trinitarios west of the Mamoré knew how to read and some voted in elections. The author continued:

> In "San Lorenzo," the Indian José Santos Noco Guaji, who exercises the functions of subcorregidor, with an office, has always maintained, under his direction, a school in which are taught not only reading and writing, but in which also are given music lessons. . . . Noted these unquestionable facts, we ask the government, Father Pierini, and the Prefect of the Missions of Tarata, can those Indians be called savages? Are they in a state of conquest those who, in the preferred place in the house-school of the subcorregidor José Santos Noco Guaji, have placed the national coat of arms, and who annually celebrate the Fatherland's anniversary with great jubilation?

The writer then described, with the concerns of a liberal, the bad effects the mission would have on commerce and on the regime of private property, and concluded by asking the national representatives from Beni to try to get the supreme resolution canceled.[49] In an equally wordy second part to the article, the same author continued in much the same vein, and listed all of the national laws, decrees, and resolutions from 1831 to 1905 which, he claimed, favored the Indians.[50]

On December 10, 1912, the three national representatives from Beni, Mariano Méndez Roca, José Antezana, and Néstor J. Otazo, wrote to the minister of war and colonization about the mission of San Lorenzo. They related how they thought the government had acted in error when it authorized the mission, saying that it had received information "of doubtful veracity." They went on to reiterate the arguments that *El Eco del Beni* had made against the mission: that the Trinitarios west of the Mamoré were not savages, that they were private landholders who attended schools, and that they exercised all the rights of citizens including the right to vote. The representatives added that the majority of those Trinitarios had, as free citizens, contracted to work on the estancias of Trinidad and its environs. "To create a mission in 'San Lorenzo,'" they told the minister, "is to snatch away an enormous quantity of hands from the agricultural and cattle-raising industry in general." The three ended by asking that the supreme resolution be declared void. On December 14, the new minister, Juan María Zalles, responded to the three representatives, telling them that he had considered their petition, and that he had requested an explanation from Father Pierini.[51]

On December 29, Pierini responded to Zalles's request. The Franciscan stated that "missions are not only founded for savages, but also for small Christendoms." He claimed that the mission of San Lorenzo would serve as a base for the conversion of Yuracarés and Sirionós, and also for the Christian and civil education of the Guayochos. He continued, saying that the Guayochos were "a tribe that are far from being the one with perfect organization and instruction that the writers of 'El Eco del Beni' and some Honorable Representatives have dreamed." Pierini described their piety as admirable, but added that their religion was full of error and pagan practices. They only worked when necessary, and only two individuals knew how to read, one being Santos Noco, "whom popular fantasy has converted into a little less than a savant." According to Pierini, Santos Noco spoke so little Spanish that he had to use an interpreter when he spoke to Bishop Santistevan in San Ignacio. The priest also enclosed one of Santos Noco's letters for the minister's perusal, sarcastically calling it "a model of literature." He pointed out that the Mojo leader and his secretary had learned to write in Trinidad, not in San Lorenzo, and that he had encountered only six students in San Lorenzo who "had conquered the alphabet."[52]

Pierini informed the minister that few Mojos owned more than four or five head of cattle, and related what Jenewein had written to him on November 13, describing how poor San Lorenzo was. He went on to say that far from being secure landholders, the majority of the Trinitarios west of the Mamoré had lost their lands to carayana patrones, and those in San Lorenzo had asked him for help in preserving theirs. Pierini concluded by saying that the Franciscans would abandon the mission if the government could show that it was not necessary in a region where the abuses of labor recruiters had caused the uprising of 1887, and where a whole tribe of people only survived because of Santos Noco. In his final sentences, the Franciscan warned Zalles "that if the Guayochos take to the woods again, or emigrate to Brazil, as they have threatened, in case the missionary is taken from them, the social and historical responsibility will not be with the missionaries, but with the writers and Representatives, who under the pretext of protecting the rights of the inhabitants of San Lorenzo, have precipitated its total ruin."[53]

Not every Franciscan supported the mission as strongly as Pierini. On December 31, 1912, Father Pesciotti wrote to the minister of war and colonization with a dissenting opinion about the mission of San Lorenzo, which was shared neither by the Venerable Discretorio nor by the editors of the Franciscan journal published in Tarata. Pesciotti said Santos Noco's people were neither savages nor neophytes, instead calling them "civilized Indians" and a "fraction of Trinitarios" (not Guayochos). He said that they had asked Pierini only for a priest to perform masses, weddings, and baptisms, not for a missionary to convert them and place them under a special mission government. He recognized that San Lorenzo was a vice cantón, and that many of its people worked for vecinos of Trinidad and its environs. Pesciotti concluded that if a Franciscan was to serve at San Lorenzo, it should not be as a missionary, but as a parish priest under the priest of Trinidad and the bishop of Santa Cruz, with a salary (*cóngrua*) provided by the state.[54]

In the end, the government permitted the Franciscans to continue serving at San Lorenzo, though it is not at all clear in what capacity. It was later called a *doctrina* or *capellanía* rather than a mission, and its priests *capellanes* (chaplains), though they never received a cóngrua. The difference in terminology was important, for missions were autonomous of all outside influence, with both civil and religious authority vested in the Franciscans. In doctrinas and capellanías, the Franciscans officially had no civil power, and were little different from parish priests.

Father Pesciotti in particular was ready to look for a way to compromise with Trinidad's carayana elite. As prefect of missions, he had written to the priest of San Lorenzo on December 12, 1912, saying "you must in no manner permit that the natives of that region, who have not satisfied their obligations to their patrones, take refuge in that settlement as a way to free themselves from their promises." Because of Pesciotti's recognition that the Mojos' debts should be honored, *El Eco del Beni* softened its criticism of the Franciscans in an article published on January 16, 1913. The author wrote that if Pesciotti's instructions were followed, "San Lorenzo will not be the refuge of idleness, which is the most marked tendency of these indigenes." The same article is notable for a final expression of the views of the liberal elite of Trinidad:

> San Lorenzo isn't, then, a reduction of savages; they are the remnants of those who rose up during the *guayochería* against what they considered to be the unjust oppression of the whites; they are always in communication with us and with the merchants of the Sécure and Chapare; they respect the organization of the family and they have especial fondness for our usages and for the practices and devotions of Catholic worship. Thus the mission of the priest there must be limited principally to instructing the town, to teaching the Spanish language in the school and the workshop, and with the language, love of the fatherland

> and of our institutions, and to contain when possible the wantonness of their customs . . . it is possible that this simple program could make the *guayochos* our best friends in fifteen years.[55]

So while continuing to insist that the people of San Lorenzo were already civilized and Christian, and thus able to sell their labor as they pleased, this liberal writer came up with a reason why a Franciscan presence could be useful. That reason was to promote national unity by teaching the Guayochos Spanish and a sense of Bolivian citizenship (both of which Santos Noco, if not the majority of Lorenzanos, already possessed). The use of the Mojo language and the maintenance of such elements of mission culture as Indian cabildos, machetero dances, and heavy drinking associated with religious fiestas continued to set the Mojos apart from the carayanas. By advocating the adoption of Spanish and the end of certain customs, the author suggested that to participate equally in liberal Bolivia, the Mojos would have to give up their ethnic identity.

By the beginning of 1913, the Trinidad elite and the Franciscans had reached a compromise. The former, realizing that the national government backed the Franciscans, resigned themselves to the residence of a Franciscan priest in San Lorenzo. The latter, admitting the difficulty of defining the Guayochos as savages, agreed not to let indebted Mojos flee to San Lorenzo to escape their patrones. Santos Noco must have agreed, preferring to take care of the families who had refounded San Lorenzo rather than antagonize the carayanas by accepting more runaways. And so it was that *El Eco del Beni* published a letter from San Lorenzo on February 1, which stated: "Of the indigenes who owe money for voyages to Mr. Antonio L. Velasco, all except two are going to fulfill their commitments."[56]

The Franciscan presence in San Lorenzo has provided us with valuable descriptions of the town. On January 3, 1913, Father Jenewein and his fellow Franciscan Fray Rufino Holler wrote a lengthy and somewhat idealized report to Father Pierini, who was now guardian of the Colegio of Tarata. Jenewein and Holler described San Lorenzo's position on a high isla, with a lagoon to the west and surrounded on all sides by extensive *curiches* (year-round bodies of water). A few weeks later, one of the priests described what San Lorenzo could be like at the height of the rainy season. "We are in this town and live without novelty," he wrote, "only surrounded by water and more water and mistreated by the mosquitoes."[57]

According to Jenewein and Holler, the settlement consisted of houses, a church, a school, a priests' house, a kitchen, a pantry, a woodshop where four Guarayos worked, and another shop for the wood turner. All were thatched with either palm leaves or straw. Six houses (including that of the priests) had been recently built, and eight others were under construction. San Lorenzo was built around a plaza with a large wooden cross in the center and one at each corner, thus exactly replicating the layout of the plaza of Trinidad and physically manifesting the Lorenzanos' desire to create a new Trinidad beyond the reach of the authorities. A sixth cross stood at the end of a road which led from north to south, and the central plaza cross was adorned with a newly built fence of turned wood.[58]

The church was spacious, but simply furnished, with a simple table for an altar, where the Mojos placed their statues of saints. The tabernacle was an antique box that they greatly valued, for they had brought it from Trinidad before the Guayochería, in their flight from "Egypt." The two fathers were building a better altar for the church at the time they wrote. Their house consisted of two bedrooms and a combined living and dining room in which was a painting of the crucifixion created by Jenewein, which every Mojo who entered paused before to venerate.[59]

The two fathers went on to describe the material and moral state of San Lorenzo. Almost every family possessed a few cattle, and some had horses. The people planted enough rice, corn, tobacco, cotton, plantains, and manioc to meet their needs, but not enough to sell. Sugarcane was rare, and the land seemed unsuited to cacao and coffee. The people did not get drunk or fight, spouses loved and were faithful to one another, and the priests were treated with kindness, respect, and obedience. The Lorenzanos showed much charity and kindness to the poor, the aged, and to orphans. Four times a year they brought the orphans to the church, where the priests blessed new clothes which they gave to the children. Whenever the townspeople killed a cow, they gave pieces of meat to the poor. There were real differences in wealth among the Lorenzanos, but the wealthier families were very charitable. Rich families (those with more cattle) were able to make cheese, sell it in Trinidad, and use the money to buy clothes for their families, servants, and the poor. Each month, a rich family would cook for all the poor, who would come together with the priest at the family's house. The priest blessed the food and was given the first piece; the poor prayed the rosary for the family, ate, and took extra food home with them.[60] We may assume that the rich included Santos Noco and the members of his cabildo, and that they were trying to act as an ideal cacique and cabildo should have, for the common good of their people.

The people continued to practice their faith devoutly, as they had in Trinidad. They got up at 4:00 a.m. every day and sang a hymn to the Virgin, and before going to sleep at 8:00 p.m., they always went outside their houses to sing a hymn to the creator. They loved music, and the fathers thought that every man in town could play an instrument, either the violin, flute, or one of the native ones (bajones or caja). The Lorenzanos celebrated their fiestas with much praying, singing, and processions, as the Franciscans observed at Christmas. The residents of San Lorenzo dressed up especially to attend church or fiestas. The men, girls, and presumably boys wore white tunics with red strips on the side, necklaces with crosses, and rosaries. Women dressed in tipoys either of cloth made in San Lorenzo or bought elsewhere, and the latter could even include silk garments.[61] The Lorenzanos' dress was almost identical to that of Mojos in Trinidad some thirty years earlier, even to the women's preference for imported cloth. Regarding religious fiestas, the fathers made no mention of the bullfighting, macheteros, and heavy drinking that should have accompanied these events. Presumably they considered such aspects un-Christian in nature, and omitted them in their generally rosy report to Pierini, who had founded and championed the mission.

Jenewein and Holler also described the two other settlements included in the district of the mission of San Lorenzo. They seldom visited Rosario, as it was far away, and the many curiches between there and San Lorenzo made the trip dangerous either by horse or oxcart. This lack of contact made the people of Rosario less enthusiastic about the mission, as Fiscal Lara had noted earlier. San Francisco was more accessible, and the fathers erected a new schoolhouse there for the teacher, Matías Cueva, who could read and write. San Lorenzo had a much closer relationship with San Francisco than with Rosario. For example, almost everyone from San Lorenzo went to San Francisco to celebrate the latter town's feast day on December 3, 1912, where they witnessed the priests bless the new cross in the middle of the plaza. The fathers mentioned that Mojos from both Trinidad and Asunta had moved to San Francisco to raise cattle within the mission's territory, indicating that other Trinitarios understood the protection the Franciscans could offer and that they acted to take advantage of it.[62]

In December 1913, Pesciotti described the situation of the Trinitarios living west of the Mamoré to the minister of war and colonization. "The Trinitarios form three clusters," he wrote, listing Rosario with its 50 inhabitants, San Francisco with 180, and the "principal nucleus," San Lorenzo de los Trinitarios, with 700 people. "They have an estancia with cattle, and dedicate themselves to agriculture, harvesting rice, corn, coffee, chocolate, and good tobacco," the Franciscan continued. Rice and corn were staples, but it seems likely that the last three crops were grown for sale to the carayanas. Pesciotti noted how faithful these Trinitarios remained to the customs of the Jesuit period. They were "markedly religious," and made cotton textiles, hammocks, camijetas, and tipoys, all of which were "very fine and very white." These cotton manufactures were an important element of the Mojo mission culture which did not survive in Trinidad itself, where looms were rare and camijetas and tipoys were going out of style by the 1880s. The Trinitarios' one conspicuous vice was a "tendency to alcoholism," which Pesciotti said they learned from the carayanas, but which could disappear with education by the missionary. The missionary served the Lorenzanos without pay, only taking what his flock spontaneously gave him, and while he governed them spiritually, politically they governed themselves.[63] Pesciotti's report is valuable because it indicates the cultivation of cacao and coffee, which Jenewein and Holler failed to note, as well as alcoholism, which they presumably preferred to ignore.

During Jenewein's term as priest of the doctrina of San Lorenzo, the national government became increasingly concerned about linking lowland eastern Bolivia to the highland core of the nation. The rubber boom had proven the economic value of the Oriente, and the loss of the Acre region to Brazil in 1903 was one sign of the need for better connections within Bolivia. Beni ranchers also knew that there was a market for their cattle in the highlands, if only they could get them there. The national government ordered the opening of a road from Cochabamba to Trinidad via the Río Sécure, and on June 14, 1915, the interim prefect of Beni, José A. Saavedra, signed a contract with Father Jenewein to build a section of the road from San Lorenzo to the Río de Moleto. The priest agreed to build a five-meter-wide road within three months of July 1, sufficiently cleared of tree trunks to allow traffic by foot, horse, or cattle drove, and with a corral and shed every five leagues. Jenewein was to receive 3,750 bolivianos from the Junta Central de Caminos (Central Board of Roads) to pay for tools, food supplies, and costs. More important, the prefecture agreed to pay for the survey and demarcation of the boundaries of the lands adjudicated to the Indians of San Lorenzo, something it had neglected to do despite being ordered to in the supreme resolution of August 30, 1912.[64] As Jenewein obviously planned to use Mojo labor for this project, securing recognition of Mojo land must have been the reason that both he and Santos Noco agreed to it. The Lorenzanos built the greater part of the road, but the project was abandoned for an unknown reason.[65] There was no overland route from Cochabamba to Trinidad five years later, so the incomplete road must have quickly become overgrown and impassable.

By 1916, there appeared more evidence that life in San Lorenzo was not as utopian as Jenewein and Holler had initially described. On January 17,

José A. Saavedra, then police intendant in Trinidad, wrote to Jenewein about the sale of alcoholic beverages to the Mojos in San Lorenzo. The intendant claimed to know that great quantities of alcohol were being sent to San Lorenzo, and were sold to the Indians at "fabulous prices." Saavedra concluded by ordering the priest to prohibit the introduction of alcohol to San Lorenzo "by all legal means."[66]

Father Pesciotti, who arrived in San Lorenzo in 1916 after eighteen hours on an oxcart, elaborated on this account of alcohol sales in his report to the minister of war and colonization on December 31. He explained that several peddlers had made it their business to bring liquor to San Lorenzo, and to exchange a can of the poorest quality liquor for an ox or a pair of cattle. He also claimed that drinking sprees were frequent because the Mojos did not spend enough time working. The only industry was the weaving of cotton camijetas for the men, as the women used clothes of foreign cloth. And although Jenewein hung Saavedra's letter on the wall of his house for all to see, nobody respected the priest's authority when it came to alcohol.[67]

Pesciotti attributed many of the problems in San Lorenzo to the uncertain status of the Franciscan there, who was neither a missionary nor a parish priest. Thus the Mojos considered their priest a missionary and civil authority when convenient, and as nothing otherwise. The nearby carayana patrones did the same, attributing missionary authority to the priest when they wanted him to mediate in disputes with Mojo workers over accounts and salaries, and treating him as a mere parish priest when he tried to defend the rights of the Indians. In the latter case, the patrones said that the priest was subordinate to the corregidor, Santos Noco (the national government had elevated San Lorenzo and all other vice cantones to full cantón in 1914). We must realize that Santos Noco himself was quite happy with the priest's uncertain status. The Mojo leader wanted a Franciscan presence in San Lorenzo to help keep carayana ranchers at a safe distance but preferred not to give up his civil authority by submitting to a complete mission regime. Pesciotti suggested that the minister resolve the anomalous situation by naming the priest of San Lorenzo both corregidor and alcalde, and paying him a céngrua and a salary as schoolmaster.[68] The minister never acted on this suggestion, which no doubt was a relief to both the carayana patrones and to Santos Noco.

While Jenewein served in San Lorenzo, the Mojos laid out a new street to enter the town, and reroofed and adorned the church, with the father himself painting the decorations. On February 7, 1919, all of this work was obliterated by a fire which swept through and destroyed the entire town, including the church, the school, the parish house, and the houses of the Mojos.[69] In the same year, the Franciscans published a report on the state of San Lorenzo and its "annex" San Francisco. There were 1,685 people in the two places, including 584 "souls" (representing married adults), 27 widows or widowers, 144 schoolchildren, and 930 children younger than seven years old.[70]

By 1919, Jenewein had been replaced by another Franciscan, Father Pedro Celestino Herritsch, also an Austrian. After the fire of that year, the mission's progress was set back, and the Mojos became somewhat dispirited. Even so, they continued to raise their cattle and to weave cloth, hammocks, ponchos, and camijetas in the mission style. In early 1919, Herritsch contracted influenza during the pandemic of that year, and then developed pneumonia. Gravely ill, he sent for the nearest Franciscan, another Austrian named Father Fulgencio Lasinger, who was at the Yuracaré mission of San Antonio del Chapare. By the time Lasinger reached San Lorenzo, it was too late, for Herritsch had died on April 13.[71] The Franciscan chronicles do not mention whether many Mojos died of influenza, because Herritsch proved to be the last missionary at San Lorenzo. In 1919, the Catholic Church removed the town from Franciscan jurisdiction (and protection) when it created the secular Apostolic Vicariate of Beni.[72]

## San Lorenzo and Colonel Federico Román

San Lorenzo did not fade back into obscurity for long, however, as it soon became caught up in a new road project of the Bolivian government. In 1921, the government sent the Bolivian army colonel Federico Román, a former rubber patrón and veteran of the Acre War, to open a new route from Cochabamba to Trinidad. His route passed through San Lorenzo, where he presented Santos Noco with a military uniform and the rank of Capitán de Salvajes (Captain of Savages) under the authority of President Bautista Saavedra.[73] Thus the national government continued to identify Santos Noco and his recently named tribe of Guayochos as bárbaros. The Mojo leader wore his new uniform and no doubt accepted his unusual military rank because identification as a bárbaro had served him well in the Franciscan period, gaining his people a powerful ally to play against the local carayana elite.

The result of Román's contact with Santos Noco was to further open San Lorenzo to the outside world. While no one had been able to visit the town during the 1900 census, the government was able to prepare complete records of property ownership there in 1924–25. The resulting catastro recognized San Lorenzo as an "indigenous community established by the Supreme Government," as well as a full cantón of the Province of Cercado. The official report identified Santos Noco as indigenous corregidor of the cantón, and gave him titles normally reserved for the carayana elite, calling him a "vecino notable" and using "don" before his name. It also revealed that Santos Noco had allowed one carayana to settle in the cantón: don Tomás Paz, who bore the title Preceptor Fiscal and who was the only other vecino notable besides Santos Noco himself.[74]

The catastro also reveals important changes wrought by Colonel Román in San Lorenzo. Of the 3,035 cattle listed, 2,006 belonged to twenty-five different Mojo owners, while 1,029 (over one-third the total) belonged to Román himself.[75] By this time, cattle represented the major wealth of the region, and the size of Román's herd compares favorably with those of the larger estancias around Trinidad. It appears that Román used his relationship with Santos Noco to gain access to land and cattle. The colonel certainly realized the value of lowland cattle to the highland market when he advocated, in 1922, the construction of a railway along the route he had opened from Cochabamba to the lowlands.[76] Román did not live in San Lorenzo, but left a Cruceño named Mariano Palacios as administrator of his two estancias, which bore the military names "Post Number One" and "Post Number Three." These properties, which had been adjudicated to Román by the national government, shared a boundary with the community lands of San Lorenzo.[77]

Why did Santos Noco allow Román to establish himself in San Lorenzo after keeping all carayanas except the Franciscans out for well over twenty years? The Mojo leader was now over seventy years old, and after losing the protection provided by the Franciscan presence, he was desperate for another ally who could help preserve San Lorenzo from the land-hungry carayana elite. Colonel Román, while not the perfect choice, had several qualities that might have recommended him to Santos Noco. He came from the Yungas region of La Paz, and hence was not a member of any of the Cruceño families that now dominated both the plains of Beni and the northern rubber forests. Though he had served with Nicolás Suárez in the Acre War, he criticized Suárez's conduct and was no friend of the most powerful family in eastern Bolivia. Perhaps most important, Román was an army officer, and thus represented the national government rather than the local Beni elite. After seeing how the national government had replaced Daniel Suárez with José María Urdininea, and then had backed the establishment of a mission at San Lorenzo, Santos Noco must have realized what a powerful counterweight it could be to the local elite.

Shortly after beginning this new alliance with Román, José Santos Noco Guaji died in 1926.[78] Santos Noco's leadership of a large number of Trinitarios for over thirty years was perhaps the most obvious legacy of Andrés Guayocho's movement. He managed to keep his people from becoming peons on carayana-owned ranches, and only sent them to work as boatmen for people he trusted, all the while conserving the form of Catholicism that the Mojos had developed in partnership with the Jesuits. Even so, he did not try to revive the way of life of the old missions and even used liberal ideology to defend the rights of his people. Through the first decade of the twentieth century, Santos Noco was secure enough in his position to treat prefects, Franciscans, and even the president of Bolivia as equals. By about 1910, however, he was getting old and was faced with an increasing number of land encroachments by carayanas, and turned first to the Franciscans, and then to Colonel Román for help. After Santos Noco's death, San Lorenzo, the messiah's town, would become just another community of workers in a landscape of large carayana ranches.

## Conclusion

Santos Noco's leadership of San Lorenzo demonstrates a number of strategies that the Mojos used to resist carayana domination. Giving up dispersion as a form of agency shortly after the suppression of the Guayochería, Santos Noco and his people returned to settle permanently at the site of the millenarian movement. They accepted his authority because he continued traditional cultural practices and managed relations with the outside world to their benefit. Although Santos Noco long kept carayanas from settling in San Lorenzo, he never tried to seal the community off from the outside world. He permitted carayanas to trade with the Lorenzanos, while they went to Trinidad and beyond to buy and sell goods, and to work for carayanas. He corresponded with prefects, priests, and newspapers in Trinidad, both to express his own views and to learn about happenings in the rest of Bolivia. He used the language of liberalism both to claim its promised equal rights under the law, and to defend Mojo lands, customs, and traditions. In the last fifteen years of his life, he maintained San Lorenzo's autonomy by deftly playing off the local elite, the national government, and the Franciscans against one another. He manipulated the dominant society's categories of civilized and savage, insisting that the

Trinidad elite accept the Lorenzanos as civilized, equal citizens, while letting the national government define them as savages when it could bring some benefit. Santos Noco's agency was based above all on participation in and engagement with the dominant society, and his skill enabled a portion of the Trinitarios to continue to live autonomously for almost forty years after the death of Andrés Guayocho.

CHAPTER SIX

# *Trinidad and San Ignacio*

## The Mission Towns after the Guayochería

For those Mojos who had remained in the mission towns, the decades between 1890 and 1930 were a time of fewer options than had existed during the early rubber boom. In this era, non-Indian immigration to Beni continued, as did the exploitation of the northern rubber forests. Carayanas secured their dominance of both the savannas and the rubber forests, and through sheer numbers could control the Indian population to a much greater degree than before. Although the Mojos found themselves in a difficult situation, they still found ways to exercise agency. The history of the Mojos living in the four former mission towns under carayana rule (especially in the two largest towns, Trinidad and San Ignacio) serves as a counterpoint to the history of Santos Noco's San Lorenzo.

At first, Mojo life in Trinidad and its environs returned to much the same as before the Guayochería. The carayana authorities trusted some of the more prominent Mojos with arms by 1889, proving how wrong Daniel Suárez had been to include the Mojo elite of Trinidad among the rebels. Several of them, including Prudencio Nosa, served as soldiers in a local column of the national army. Although these Mojos may have been empowered through membership in the military, they were outnumbered by carayanas, they were all of the lowest rank, and they were invariably listed last on the rosters. This reflects both carayana opinions of their place in society, and the Mojos' inability to change those opinions.[1]

A few years later, some Mojos learned that membership in the military could expose them to the very hazards they sought to escape. In 1892, Sub-lieutenant Quintín González Portal made a contract with rubber patrones on the Río Madre de Dios, agreeing to provide them with workers. As his father, Dr. Samuel González Portal, was prefect of Beni, Quintín had no trouble recruiting workers. The prefect sent soldiers from the column to the rubber zone, ostensibly to garrison Villa Bella at the confluence of the Beni and Mamoré rivers, but really to gather rubber on the Madre de Dios. The prefect and his son also took peons from ranches around Trinidad and tried to recruit free Mojos from the environs of the capital. The latter fled their fields to escape impressment, and the envoys of the prefect contented themselves with robbing all of the Mojos' food and provisions. On January 6, 1893, the González Portals even ordered their soldiers to rob the offerings from an altar to the infant Jesus that the Mojos had set up in the atrium of the Trinidad church.[2] But despite the González Portals' abuses, several factors combined in the 1890s to lessen the Mojos' participation in rubber exploitation in the northern forests.

First, the demand for boatmen decreased. By the early 1890s, regular steamboat service was established above the cachuelas on the Mamoré, Iténez, and Beni river systems. Bolivian boatmen continued to handle traffic through the cachuelas before the completion of the Madeira-Mamoré Railroad in 1912, and Mojos in particular continued to ply the river routes above Trinidad toward Cochabamba and Santa Cruz, where steamboats were impractical. But from 1892, steamboats operated on the Mamoré between Trinidad and Guayaramerín, so the need for Mojo paddlers decreased after this date. Because of their knowledge of the river, many Mojos served as pilots on the new steamboats.[3] The spread of steamboats freed the Mojos from one kind of labor, but it brought an end to the days when a large crew of physically strong paddlers could dominate their outnumbered merchant patrones and removed an opportunity to work for cash in an enterprise the ex-mission Indians once dominated.

Second, the Bolivian government began to create new protective legislation. In 1896, the national government passed the *Ley de enganche* to try to regulate the recruitment of labor for extractive industry (principally rubber), agriculture, and commerce north of the fourteenth parallel of south latitude. Henceforth, patrones were to stipulate peons' salaries and the duration of the contract beforehand, they were to pay for the peons' travel both to the rubber zone and home again, and they were to provide sufficient food and medical attention for their workers. In addition, they were to keep peons informed of their accounts, were not to transfer workers to other patrones without the workers' consent, and were not to give advances of more than four hundred bolivianos. The contracts were to be made before the local authorities, and patrones were to deposit two hundred bolivianos to ensure the peons' return, which would go to their heirs if they died in the rubber zone.[4] In a decree of 1897, the government further clarified the Ley de enganche by specifically stating that it applied to recruitment in the four Mojo towns, as they were below the fourteenth parallel. This decree also forbade whipping and the stocks as methods of disciplining workers in the rubber barracas.[5]

In lands so far away from the national government, little could be done to enforce these laws, and whipping continued to be used as a punishment.[6] Even so, the laws must have had some effect, as they provoked protest, especially in Santa Cruz.[7] Some Trinidad patrones made efforts to comply with the provisions of the Ley de enganche, as when one offered a good salary and "sufficient food and remedies, for free," to prospective crewmen, given proof of the peon's freedom from other contracts.[8] But abuses continued as well, as when the Trinidad police gave peons advances so that they could be declared free of debt, and then sent them to work for two rubber patrones on the Río Orthon.[9]

Third, patrones found new sources of labor for rubber extraction, from Brazil or even further afield. Another drought in Ceará sent more northeastern Brazilians into the region in 1899, and many Brazilians settled in Bolivian territory. Brazilians so dominated Acre that Bolivia lost it to Brazil in 1903, and Brazilians settled even further south. Only one Bolivian lived along the Abuná River (the new international boundary) in 1907, while Brazilians gathered rubber on both banks and sold it to Brazilian river traders. A Bolivian report on the area between the Manuripi, Orthon, and Beni rivers and the new border with Brazil estimated the population as 35 percent Brazilian, 25 percent Bolivian, 10 percent Peruvian, 8 percent Syrian, 5 percent European, 2 percent Japanese, and 15 percent other in 1918–20.[10]

A fourth factor is that carayanas needed Mojos as laborers in the environs of Trinidad and did not wish to send them north. In the greater part of Beni, which lay to the south of the rubber zone, the period from 1890 to 1930 was characterized by a rapid growth of non-Indian landholding as more and more carayanas turned to cattle ranching. A few also established sugar plantations (called *establecimientos*) around San Ignacio at this time.[11] The growth of the cattle industry on the savannas of Beni was directly linked to the rubber boom. The resettlement of large areas of northern Bolivia and adjacent parts of Brazil by rubber tappers created a market first for jerky and then for live cattle.

By the first decade of the twentieth century, steamboats were being used to transport cattle alongside human passengers. But as steamboats' freight costs were high, cattle merchants would more commonly float thirty to thirty-five head down the Mamoré in *yangadas* made of two batelones joined together. At Guayaramerín above the first cachuela, they would disembark their cattle and drive them overland to Riberalta, Cachuela Esperanza, and Villa Bella. By 1904, cattle merchants had opened an overland route from Exaltación to Riberalta.[12] By 1906, Beni ranchers were exporting so many cattle to the Brazilian Acre that the Bolivian national government decided to levy a tax on each animal sent there. Ranchers along the Río Yacuma from Santa Ana to Reyes, from Exaltación, and even from San Ignacio sold cattle in Acre as well as in the Bolivian Territorio Nacional de Colonias (the modern Department of Pando) and Riberalta.[13] Ranchers in the Mojo area sold most of their cattle in Santa Cruz, Chiquitos, and Argentina, and this exportation was continuous from the early 1890s on.[14] One writer claimed that the Beni cattle industry would go into decline because too many cattle were being exported to either Santa Cruz or Acre.[15]

## Land Tenure

Though created to provide land for all Bolivians, the Ley de tierras of 1905 provided the framework through which individuals could acquire large portions of land, which on the savannas of Beni would be used for cattle raising. Land petitions increased following the collapse of rubber prices, as carayanas increasingly turned to ranching.[16] In many cases, carayanas acquired larger properties than technically permitted under the 1905 land law. They evaded provisions intended to prohibit the formation of latifundia by registering several contiguous properties, whose consolidation formed extensive ranches.[17]

Zulema Lehm has compiled statistics using Trinidad's Archivo de Derechos Reales that reveal several trends about land transactions in the Province of Cercado in this period. In 1895, ninety-two transactions were registered, but for the next five years there were only between one and fifteen per year. The total number of transactions per year jumped to 205 in 1905 and reached a high of 748 in 1915, remaining at over 500 per year from 1910 to 1930. The percentage of total transactions representing transfers from Indians to carayanas generally declined in the 1895–1933 period, but the number of property sales from Indians to carayanas reached highs of eighty-nine in 1910 and seventy-three in 1915 before falling in the 1920s and 1930s. The years after the land law of 1905 and the collapse of the rubber boom in 1912 were thus characterized by a large number of transactions, and by a loss of Indian land to carayanas.[18]

Many Mojos lost land to carayanas through inability to participate in the Bolivian legal system. Indians who did not register their properties with the state may well have lost them to unscrupulous carayanas who petitioned for their lands. By 1912, a newspaper in La Paz reported that the situation of Indian landholders in Beni was "truly critical." According to the writer: "Some petitioners for tierras baldías, supported by acts of concession from the prefects, try to snatch those lands held of old [by Indians] . . . many of those indigenes have very old property titles, but because of their social and economic condition they are not in the position to assert their rights before the departmental authorities, much more so if their opponents live in the seat of the prefecture."[19] Despite one incident when Prefect Fabián Vaca Chávez refused accountant Clodimiro Rodal's petition in the cantón of San Pedro because the land was already occupied by Canichanas, the archives offer evidence supporting the above newspaper account.[20] Such was the case with a carayana petition that was confirmed, despite the fact that the land in question, south of Santa Ana, was occupied by an unknown person's cattle.[21]

However, Mojos were not completely shut out of the land market. Lehm's data show that Indians received state and municipal land grants, bought land from carayanas, and bought and sold land among themselves in the 1905–33 period.[22] Trinidad newspapers of the era published a few Mojo land petitions, including two near Trinidad, and one to the north of San Javier.[23] Land petitions of both carayanas and Mojos occasionally mention adjacent Mojo properties in the description of their boundaries. At least one Mojo, Valerio Tamo, employed the carayana tactic of petitioning for contiguous properties in order to control more land.[24] Nevertheless, Mojo properties were generally fewer and smaller than carayana ones. The Indians also tended to concentrate on agriculture on their lands, while carayanas focused on cattle. Some Indian properties were located quite close to the Mamoré or its old meander cutoffs, and thus were especially exposed to flooding and a loss of their food crops.[25]

Perhaps the most interesting evidence of Mojo landownership is that which indicates their efforts to preserve communal property. In 1926, five Trinitarios published an advisory affirming that none of them had the right to sell any part of the property called San Antonio, which belonged to the parcialidad of Tejedores, and had been confirmed to that group by order of the prefect in 1912. In 1913, seven other Mojos petitioned for communal title to a single piece of land, a five thousand-hectare estancia southeast of Trinidad named Santa Rosa. With the approval of carayana officials and the Mojo corregidor of the Trinidad cabildo, they received title to the land. Thus some eighty years after carayana authorities had first mandated private property in Beni, at least two groups of Mojos managed to retain communal title to land. These documents also prove that, despite a marked decline in their power, both the cabildo and the parcialidades survived and continued to play a role in Mojo society, as they still do today.[26]

## Labor

The new carayana ranches and sugarcane plantations needed labor, and they acquired it in the form of peons who were generally called *mozos* (servants) or *colonos* (tenant farmers). In the Province of Cercado, which included the Mojo and Canichana towns, most of the mozos were Indian (especially in San Ignacio), though there were also many Spanish-surnamed peons (mestizos of local or Cruceño origin) among them. The Indian peons included a good many people who had lost their land to carayana petitioners.

The carayana patrones tried to control their mozos through debt and through the *matrícula de mozos* (registry of servants). It is unclear exactly when the matrícula was created, but Lehm, Lijerón, and Vare speculate that it evolved out of an earlier registry of tributaries, and represented a "privatization" of previous indigenous-state relations. This is indeed possible, for the national government had ended Indian tribute in Beni in 1883 and in the rest of the nation in 1885. Although tribute continued as a *contribución territorial de indígenas* in most departments into the early 1900s, Beni never created a departmental-level tribute to replace the national levy (it was the only department without one in 1900). The matrícula was officially sanctioned by the state during the rule of the Liberal Party (1899–1920) and continued under the ensuing Republican Party regime (1921–34), despite supposedly being suspended.[27] In the matrícula, a worker and a patrón would go before a local authority, either the police intendant in Trinidad or the corregidor in San Ignacio. There the worker would be registered to the patrón in a legal document in an attempt to secure a stable labor force. One Mojo's registry used a form typical of all: "August 21, [1913]. Eugenio Jou, registered by Dr. Francisco P. Vargas, for the compulsory term of three years and with a salary of thirty bolivianos [per month], and in case of desisting before the contract expires, will pay the fine of 100 bolivianos." Another contract stipulated that if the mozo ended his term of service in debt, he would have to work another year for his patrón.[28] Although there are indications that carayanas wanted every Indian to be assigned to one patrón or another, some Indian commoners resisted being drawn into such contractual obligations. They feared that after accepting an advance and agreeing to a contract, their patrón would send them to work in the rubber zone, a fate worse than peonage close to home. As a result, at least some Mojos remained free, which in one carayana's opinion meant that "they devote themselves to rest and drink."[29]

A carayana wrote in 1913 that mozos in San Ignacio, "as in all Mojos [i.e., the Province of Cercado], are born and die in the power of the same family"; moreover, they received a miserable salary and were charged for their agricultural tools by their patrones.[30] Other reports indicate that mozos did not spend their whole lives bound to the same family. Another carayana remarked on the "instability of the peonate" in 1913, and described how, despite all laws and regulations, many mozos changed patrones as easily as they changed shirts.[31] Mozos did flee of their own accord, but sometimes they were aided by carayanas. Carayana patrones were extremely competitive over resources ranging from cattle to the labor of mozos and often attempted to recruit each other's laborers, a behavior that the matrícula was supposed to curb. At times this competition could have favored the mozos, but at other times they were essentially kidnapped by one patrón from another.[32]

It is possible, however, that Indians were less likely than mestizos to move from patrón to patrón. One contemporary complained that Beni patrones who recruited Cruceño mestizos ended up conscripting "the peon who is disposed to swindle as many patrones as have faith in him."[33] Anthropologist Dwight Heath described such an extreme willingness to change patrones among Santa Cruz mestizos in the 1950s. As one outcome of such migratory lives, Cruceño peons formed only brittle consensual unions, which they broke every time they moved to a new location, leaving women to head the household until they accepted a new man as partner.[34] In earlier chapters, we have frequently noted the religiosity of the Mojos, and the high esteem in which they held marriage. Mojo men were probably less likely than Cruceño mestizos to break up their families and move to a different estancia. We cannot assume that the pressures of peonage markedly altered Mojo attitudes toward marriage and family in these decades. As recently as the 1990s, Bolivian statistics show that 22 percent of households in the Department of Santa Cruz were headed by women, while only 6 percent of Mojo households had female heads.[35]

Mojo oral traditions remember the time of the matrícula as an unhappy period of their history. Elders in the 1990s called it the "time of slavery," and remembered how the patrones dragged Mojos before the authorities to register them as peons. The matrícula established a fixed term of employment, but the term never came to an end. The mozos had to perform their tasks on the labor-intensive sugar establecimientos, and cultivate their own chacos as well. If mozos tried to escape and were captured, the patrones would have them whipped or have cuts made on their buttocks, and would rub salt in the wounds. Patrones were supposed to give mozos clothing and food, and they gave their workers a few days' freedom around the feast day. At that time, they would provide their mozos with abundant alcohol, letting them drink during the fiesta as long as they returned to work afterward. Mozos had nowhere to turn for help, as the white authorities, the priests, and the indigenous cabildo all participated in their exploitation or turned a blind eye to it.[36]

According to Lehm, Lijerón, and Vare, each establecimiento had its own patron saint, and the Indian mozos celebrated that saint's fiesta at the establecimiento, following their traditional ceremonial practices. Thus the establecimientos, like the free refugee communities examined in the last chapter, contributed to the decentralization of ceremonial authority, which was no longer controlled by the cabildos of the four former mission towns. Many of the establecimientos thus became small Indian communities in their own right.[37] According to oral history collected by James Jones, patrones and mozos around San Ignacio all went to town to attend the celebrations at Christmas, Holy Week, and the patron saint's day on July 31. Patrones also gave mozos one day off per week, usually Sunday.[38]

Patrones, living on isolated properties with large numbers of mozos, were occasionally given reason to fear their workers. In 1908, Rómulo Suárez was killed by his *criados* (servants raised in the household as children) at his estancia, La Loma, near Trinidad. It was said that he treated his people with extreme strictness, that "he had a penitentiary in the form of his house," and that he had threatened his seventeen-year-old criado Cicerón with five hundred lashes. Cicerón and another servant then killed Suárez one night, hitting his head with an axe and a large wooden pestle. The other servants altered the crime scene to make it appear as if Suárez had fallen from the second story of his house, and then mourned him at his wake and burial as if it had been an accident. Rómulo Suárez may have been the most famous patrón to have been killed by his servants, but he was not the only one. Before this, Indian servant women had killed their patrón Octavio Roca, and later in 1908, the patrón Copertino Ortíz was killed by one of his mozos. The murder of patrones by mozos was one of the carayanas' worst fears, for it reminded them (as did the memory of the Guayochería), that Indians and mozos could employ violence when pushed too far.[39]

## Population

The Bolivian census of 1900 provides us with much information on the four Mojo towns at the height of the rubber boom (see table 6.1). Its statistics tell us that Mojos were a minority in Trinidad and Loreto by then, owing to the large immigrant population that settled in those towns. San Ignacio, and to a lesser degree San Javier, still had Mojo majorities. In all four towns, Mojo women outnumbered Mojo men, sometimes by a wide margin, presumably because of the loss of males in the rubber zone. In contrast, men always outnumbered women among whites, and almost always among mestizos, reflecting the recent immigration of white fortune-seekers and mestizo workers to the area.

## San Ignacio

The history of San Ignacio in this period was significantly different than that of Trinidad and the other towns near the Mamoré. Most of the difference was due to San Ignacio's distance from the major river networks, as it

**Table 6.1** Population of the four Mojo towns in 1900, by race and sex (racial categories "negra" and "no consta" omitted)

| | *Trinidad* | *SanIgnacio* | *Loreto* | *San Javier* |
|---|---|---|---|---|
| Total | 4,294 | 2,265 | 1,649 | 383 |
| Indigenous males | 764 | 670 | 326 | 92 |
| Indigenous females | 1,001 | 801 | 375 | 122 |
| Mestizos | 768 | 113 | 119 | 31 |
| Mestizas | 668 | 85 | 70 | 36 |
| White males | 448 | 163 | 398 | 46 |
| White females | 394 | 121 | 260 | 23 |
| % Indigenous | 41% | 65% | 43% | 56% |

*Source*: Oficina Nacional de Inmigración, Estadística y Propaganda Geográfica, *Censo general de la población de la república de Bolivia según el empadronamiento de 1° de septiembre de 1900. Tomo II: Resultados definitivos* (Cochabamba: Editorial Canelas, 1973), Censo de la población en 1900, 21.

was four days' travel from Trinidad by oxcart in the dry season. Carayanas recruited some Ignacianos for rubber work, but not nearly as many as in Trinidad and other river towns. Before the 1890s, San Ignacio's only resident carayanas were the corregidor, the priest, and very few hacendados (including Nemesio Saavedra and Félix Montenegro in 1887). By 1891, when Saavedra rigged his own election to the municipal council, and 1892, when his second election was declared void by his rivals, there were enough carayana patrones to engage in local power struggles. A few more whites arrived later in the decade, but James Jones notes that no white births were recorded in the parish archives until 1900. After that date, there was a general increase in white births, from 3 in 1900, to 17 in 1905, 16 in 1910, 28 in 1915, 86 in 1920, 71 in 1925, and 110 in 1930. Whites began to serve as godparents to Indians from 1910 on, but only at a miniscule percentage of Indian births.[40]

San Ignacio was considered an important agricultural and stock raising center in the early twentieth century. The basis of its prosperity was its dense indigenous population, since it did not lose as many of its inhabitants to labor recruitment for rubber exploitation as did Trinidad. At the time of Prefect Vaca Chávez's visit in 1912, San Ignacio still had an abundant indigenous population. Carayanas considered the Ignacianos to be the opposite of the Trinitarios in two aspects: the Ignacianos were "laborious, docile, and tractable," and were more likely to assimilate and adopt white customs. The cantón of San Ignacio was known as "the land of cacao and tobacco" for two of its principal agricultural products. Orbigny noted immense groves of cacao at San Ignacio in 1832, and that crop continued to flourish there in 1912, when the Ignacianos were reputed to be the only former mission Indians who still cultivated it. San Ignacio's estancias supported at least thirty thousand head of cattle, and the cantón provisioned Trinidad with food.[41] The catastro of 1924 listed many more "citizens" (carayanas) than Indians as property holders in San Ignacio, even though Indians were still more numerous than whites there.[42]

Statistics that show a carayana dominance of property-holding in a Mojo-majority cantón reflect the growth of peonage. Jones, relying on oral testimonies, estimates that 40 to 60 percent of Ignacianos were mozos in 1930. This means that a roughly equal number of Ignacianos (some 1,032 to 1,548 according to Jones) were not bound to patrones at that time. These free Ignacianos were wealthier than the mozos, and had houses in town and gardens within three or four kilometers' distance. Almost all owned coffee and cacao groves, cattle, oxen, and horses. Some also planted sugarcane and had sugar mills. They also occupied the most important cabildo posts.[43]

Official statistics portray far fewer Ignaciano property holders than Jones's account would suggest, though perhaps not all of them were registered. The national government recorded twenty-one Ignaciano landholders in 1902 and most of the same names appear in a list of an equal number of Ignacianos who owned cattle in 1901 and 1902. Two Mojo landowners even had mozos, one with four and another with ten. The number of cattle owned by individual Mojos varied greatly, from five head to four hundred head. As in other cantones, the richest people in town were carayanas: Avelino Rivero had sixty mozos and eleven hundred cattle, and María de la O Cardoso owned twenty-one hundred cattle. Even so, more Ignacianos owned land and cattle in 1902 than did the Mojos in any other of the former mission towns. In that year, no Trinitarios owned (or had legally registered) rural property, and only nineteen owned cattle. Records from 1925 also show more Mojo property holders in San Ignacio than the other towns. In that year, no Mojos were registered as rural landowners in Trinidad, while six were in San Ignacio; three Mojos were registered as cattle owners in Trinidad, while at least twenty-three were in San Ignacio.[44] Throughout this period, then, Mojos in San Ignacio maintained more autonomy than did those in Trinidad, though neither group had as much control over their lives as did the Mojos of San Lorenzo and San Francisco.

Because of the different historical experience of San Ignacio, the Ignacianos managed to maintain more of their culture and traditions than Mojos in Trinidad and the other towns. According to Jones, Ignacianos continued to wear the camijeta and tipoy in 1930, long after they had gone out of style in Trinidad. They also conserved the Mojo language to the point that few people could speak Spanish in 1930. Jones found that belief in spirits related to nature was still strong in the 1970s, and Lehm observed evidence of belief in ancestor spirits somewhat later. Jones also describes the continued existence of the cabildo, músicos, sacristanes, abadesas, and macheteros and their importance in ceremonial life. The strength with which these socioreligious organizations have survived has earned San Ignacio the reputation of folkloric capital of Beni.[45]

There are, however, few accounts of daily life in early twentieth-century San Ignacio. One corregidor's despotic rule generated written protests that reveal much about internal conflict, both between whites and Indians, and between patrones. In October 1905, Corregidor Nicanor Gil abused his authority to appoint and replace the Ignaciano cacique. When Cacique Manuel Jare wanted to resign his post, Gil told him he would accept only if Jare worked on some boats he was building for an extremely low remuneration. The corregidor then forced the Ignacianos to spend a month building corrals for him, paying each worker four bolivianos. The Indians lost time for their own farm work, and at least one was unable to mill his own sugarcane. At his own estate, described as a "penitentiary," Gil put the mozo Teodoro Yucu in the stocks for three nights. He then gave Yucu to the patrón Leandro Gutiérrez, even though his real patrón was Fenelón Suárez.[46]

In December 1905, Gil disturbed the peace of San Ignacio by riding out with a group of men to avenge himself on his local enemies. First, they broke into the house of Inocencio Nálema, a Mojo who had sued Gil for unspecified crimes (a significant example of Mojo use of the carayana legal system). They tied up Nálema and his son Pedro, and then went to the house of Fenelón Suárez, who had also quarreled with Gil. They threatened Suárez, and Gil tied up the Mojo mozo Mateo Beyuma and carried him away, claiming that Beyuma was in his indenture even though he had been registered in the matrícula to Suárez. Gil took the Mojos and one other employee of Suárez to his own establecimiento, and Suárez's lawyer later opined: "Their fate is unknown; if they have not yet been murdered, they have at least been whipped and put in the stocks."[47] In his last year as corregidor (1908), Gil continued his abuses, kidnapping another mozo and fencing in the urban property of the Mojo Juan B. Nálema for his own use. His replacement's principal task was to guarantee the patrones control of their mozos without interference, thus explaining the real reason Gil had been considered a despot.[48]

## New Regulations

Turning to Trinidad, there are abundant sources that we can use to analyze conditions in the departmental capital in the early twentieth century. A report by the national government stated that the town of Trinidad had a population of 2,556 in 1901. The town was composed of between ten and twelve blocks, none of which was completely built up. The few manufactures it produced included sugar, cigars, and hammocks, the last being the only survival of the once thriving textile industry. Cattle were the only local source of wealth, and carayanas still went to the Río Beni to seek their fortune.[49]

As non-Indians grew secure in their dominance of the former mission towns, so did carayana forms of government and bureaucracy mature. The carayanas' governance of the former mission towns in the first decades of the twentieth century was marked by a new zeal for regulation. Though this trend led officials to demand that oxcarts and even puppies be officially registered, there were several other policies which had a serious impact on the Mojos' lives.[50] The police in Trinidad seem to have increased their activity in this period, and Trinidad newspapers frequently published lists of the people that had been arrested or fined. Many, though not all, of the arrestees were Mojos, and their offenses were usually drunkenness and fighting.[51] One carayana suggested that "those detained for drunkenness and other faults" be put to work digging drainage canals around Trinidad, and the authorities may indeed have forced arrestees to provide free labor.[52]

The carayana press applauded one police intendant, José A. Saavedra, for his measures against alcoholism and vagrancy, which affected both Mojos and poor carayanas. The intendant prohibited the sale of alcohol to policemen, minors, and to Indians, mozos, and servants on workdays.[53] Saavedra, like most carayanas, was thus only concerned that Indians, mozos, and servants be fit to work; whether they abused alcohol or not in their free time was unimportant. One carayana writer lamented this lack of concern when a Mojo mozo named Vicente Tamo was discovered dead one morning in 1912, after drinking excessively, and complained that drinking was widespread among mozos and servants.[54]

One senses that the carayana elite was deeply concerned with social control and with "civilizing" the streets of Trinidad. Their controls were only partially effective, however. One carayana writer pleaded for the police to fine "stupid" oxcart drivers who put pedestrians at risk with their half-tamed oxen. "On more than one occasion we have seen ladies and gentlemen doing ridiculous pirouettes crossing the street, in order to reach the sidewalk and save themselves from being struck by oxcarts," the writer observed.[55] It is not hard to imagine an Indian or poor mestizo carter exercising a covert, "everyday" form of resistance in pretending to be unable to control their oxen.

Another aspect of the carayana bureaucracy which affected the Mojos was the *impuesto de prestación vial*, a tax (created in 1905) to support road work or canal building that could be paid in either money or through actual labor on those projects.[56] In 1913, a Trinidad newspaper listed the names of those who had paid the required six bolivianos and of those who had paid in labor service. Of the 524 men in the former group, only 24 had Mojo surnames. Of 207 men in the latter group, 116 had Mojo surnames.[57] Thus, although the tax did not specifically single out Indians for road and canal work, they nevertheless ended up providing the largest number of laborers. The impuesto de prestación vial was in effect a disguised form of forced labor that the carayana elite imposed on the majority of Mojos and on the poorest of the non-Indians resident in Trinidad.

The prestación vial had its greatest impact around Trinidad, where the departmental authorities' rule was most effective. Elsewhere, such labor for the state was probably resisted as much by patrones who resented attempts to diminish their own workforce as by the prospective laborers themselves. In those parts of the countryside where the government's mandate was respected, not even mozos were exempt. In 1913, for example, the authorities decided to cut a channel across one of the Mamoré's meanders at a place called El Corte de Santa Catalina. The contributors to the prestación vial included sixty-five workers of three different patrones, and twenty-nine independent persons. All except five of the free men paid with labor rather than money, and those five were all carayanas, two of them related to two of the patrones. At least thirty of the laborers were Mojos, and most of them worked for a single patrón.[58]

The Mojos faced another burden when the national government decided to reform the military. The Liberal Party that took power in the revolution of 1899 was intent on reorganizing and modernizing the armed forces, and began a series of reforms, including a decree authorizing conscription in 1904, and culminating in an order for compulsory military service in 1907.[59] The Mojos were affected by these reforms as early as 1905, when the authorities in Trinidad required several hundred men to register for military service before choosing a few recruits at the end of the year. Although only about a third of the men who registered had Mojo surnames, two of the three ultimately conscripted were among them. In Loreto, the one recruit was a Mojo, even though most of the town's population was non-Indian. In these two towns, there seems to have been some bias against Indians in the selection process. San Ignacio's two recruits, however, both had carayana names, even though the town was almost two-thirds Mojo.[60] In 1906, recruitment followed a similar pattern: two of three recruits from Trinidad were Mojos, as were the one from Loreto and one of two from San Ignacio.[61]

Ironically, military registration, which should have worked to strengthen Indians' place in the Bolivian nation, created a loophole exploited for labor recruitment. In 1906, a carayana detained paddlers in Cuatro Ojos (a river port on the route to Santa Cruz), for not carrying proof of military registration. He then forced the paddlers to work for him for two months.[62] Failing to register earned one the label of *omiso*, and carayana authorities felt entitled to capture omisos and subject them to forced labor. A law required that omisos provide six months of labor on public roads, and so they essentially were added to the prestación vial workforce. The greater part of this abuse fell on Indians and mozos, who were taken away from their fields. In 1930, a carayana writer proposed that omisos be provided to Federico Román, by then a general but still an indefatigable road builder in the Oriente.[63] There was great interest in road building after rubber prices began to deteriorate in 1912, when Beni ranchers found that their cattle could no longer command high prices in Acre. In search of a secure market, carayanas in places like Reyes, San Borja, San Ignacio, and Trinidad began trying to construct roads up the Andes to Cochabamba or La Paz, and presumably employed mozo labor.[64]

Discrimination against Mojos applied even to regulations on public discourse. This is especially clear in punishments for calling people derogatory names. To call someone a camba (a disparaging term for an incorporated lowland Indian) was punishable by four days' to four months' imprisonment.[65] But as carayanas frequently called Mojos "cambas," it seems likely that the law was only applied when someone affronted a carayana with this name. In contrast, not even high-ranking Mojos could get away with calling a carayana a derogatory name. In 1906, for example, the respected old cacique of Loreto, Manuel Muiva, asked, "How can they name a mozo like us corregidor?" Corregidor Melchór Cortéz responded by discharging Muiva as cacique, putting him in the pillory, and threatening to whip him.[66]

## Urban Property

Despite having lost most of their rural property, Mojos continued to own urban lots in Trinidad in the early twentieth century. These Mojos contracted carayana legal representatives to validate their titles, which had to be drawn up according to a formula and then published three times. In 1919, Pedro Pablo Maza contracted the *procurador de causas* Pedro C. Rojas to petition for municipal land in the center of a block some two and a half blocks from the plaza. The fact that three other Mojos owned lots in this block north of the Calle 24 de Septiembre suggests that it was still a predominantly indigenous neighborhood.[67]

Despite this evidence of continued Mojo property ownership in Trinidad, native people were slowly giving way to the carayanas and were forced to live further away from the plaza or even to leave town. As carayanas grew richer and Mojos grew poorer, the former began to buy urban properties from the Indians. White officials and merchants began to appropriate the area around the plaza for themselves as early as the 1860s and

1870s, when Corregidor Frutos Nosa and Cacique Manuel Visitación Noe sold their properties to carayanas. The Guayochería furthered the extension of carayana control in Trinidad, as many Mojos fled the town and as the authorities killed many of the cabildo leaders. Mojos continued to sell urban lots to carayanas in the 1890s, as when the widowed spinner Pabla Muiba sold her property to the vecina Estefanía Coronado for the very low price of forty pesos.[68]

The process of displacement continued in the early twentieth century. In the decades from 1900 to 1930, the prefecture gave urban lots to Indians, who ended up selling most of their properties to carayanas a short time after, a tactic carayanas used to cheaply acquire more urban land.[69] In 1907, a carayana writer noted that many Mojos had been selling their houses in Trinidad and moving away, scared off by certain whippings that had been carried out. The writer did not explain further but compared their flight to that before the Guayochería.[70] By 1914, there were only about five hundred Indians left living in Trinidad, along with five hundred whites and fifteen hundred mestizos.[71]

A similar process of displacement occurred in San Ignacio, but at a later date. In the 1890s, the Ignaciano cabildo families still resided around the plaza, and a few of these elite families still had houses on the plaza in 1930. Other free Mojos lived in San Ignacio then, though on the town's periphery. It was not until after 1930, however, that all Ignacianos were relegated to easily flooded lowlands on the edge of town, and that the plaza came to be an exclusively white space, approached by the Indians with timidity.[72]

## Education

As noted in chapter 2, education for all citizens was one of the main tenets of liberalism, and schools had existed in Beni towns since the 1840s. By 1906, however, most schools had been closed in the outlying cantones of Beni, except for those in Santa Ana, Villa Bella, and Riberalta. In Trinidad in that year, there were primary schools for girls and boys, but still no secondary school. In 1907, there were four schools in Trinidad, with no Mojo students except for Simón Semo, who attended the capital's Escuela Infantil.[73] Shortly thereafter, there is evidence of a separate school for Indians in Trinidad, though its officials were carayanas, and its teacher presumably taught in Spanish only. In 1912, one of the duties of the priest of Trinidad was to maintain an Indian school. In 1913, the Municipal

Indian School of Trinidad (presumably run by the church) counted fourteen students, all boys.[74] More doctrinaire liberals would have wanted education to be secular, but in Trinidad, the carayanas were happy to leave Indian education to the church. By 1930, the school was taken over by the Apostolic Vicariate of Beni, and was called the Parish Indian School, though it was visited by inspectors belonging to the secular government. In that year, the inspector received a special greeting when student Francisco Cayuba gave a flowery and complementary speech in Spanish on behalf of himself and the other ten boys.[75] Though there were not many students in this Mojo school, it is apparent that they were learning to participate in the carayana world, using Spanish and all of the courtesies required to impress the authorities. The Mojos must have desired education both as a way to demonstrate their equality and to provide more opportunities for their children.

## Holiday Celebrations

The Mojos continued to observe the Trinidad feast day in this period, and its bullfights and drunkenness became a contested issue in the now carayana-dominated capital. The carayanas used this central event of the Trinitario calendar to paint the Indians as barbarous and backward. In a newspaper editorial before the fiesta of 1905, a carayana called the bullfight the "last remnant of barbarism that we have," lamented the loss of work due to the fiesta, and sympathized with patrones who feared that their mozos would be killed.[76] One carayana writer predicted that the bullfight of the 1906 fiesta would be "as always: barbarous and inhumane, worthy of other times," while another in 1908 sarcastically described the sight of drunken Indians being hit or killed by bulls as a "humane, pious, and civilizing spectacle."[77] In 1912, another carayana suggested that the bullfight be replaced with greater alcohol consumption, stating: "God grant that the bullfights not be permitted, because in the environs of this capital there are neither wild bulls nor Indian bullfighters: this game has here degenerated much, turning into a brutal and tasteless hubbub of drunkards. To satisfy the indigenes, give them aguardiente, and more aguardiente, and more aguardiente."[78] However, other carayanas could support traditional aspects of the fiesta when it suited them, as when the police fined several citizens eight bolivianos each "for not having brought bulls for the fiesta as is the custom."[79] Some carayanas even appreciated the bullfights when they were well-managed by the police.[80]

The carayana citizens introduced their own yearly festival with liberal and patriotic overtones with the celebration of the *fiestas patrias* on August 6, the anniversary of Bolivian independence. In 1895, they set up a secular "altar of the Fatherland" with busts of Simón Bolívar and Antonio José de Sucre in the Trinidad church atrium, and Mojo macheteros were one of many groups that took part in the parade that marked this holiday.[81] In 1906, the fiestas patrias included a civic procession to another altar of the Fatherland, a presentation by the girls of the Bolívar school, a patriotic speech by the prefect, an orchestra, the national anthem, and a theatrical depiction "in which Bolívar presented his favorite daughter [Bolivia] to the world."[82]

Despite the obligatory machetero performances, the Trinitarios at first refused to fully participate in the spirit of this secular holiday, "as if they realized that liberty had still not arrived to benefit them." Then in 1913, they decided to take part and marched through the streets with an orchestra, singing the national anthem and acclaiming Bolivia and the local authorities. A carayana observer assumed that the Trinitarios' participation was evidence that they felt an alleviation of their servile status, and that the "indigenous class" was being incorporated into national society.[83] A more likely explanation is that the Trinidad cabildo took the decision to demonstrate Mojo loyalty to the nation as a way to demand the equal rights under the law promised by Bolivian liberalism. As even carayanas commented on the discontinuity between the freedom proclaimed at the fiestas patrias and the actual treatment of mozos, we can assume that the Mojos were well aware that proclamations of liberty and equality meant nothing unless they participated in the system themselves.[84]

## The Church

Despite their participation in the fiestas patrias, the Mojos remained most attached to their religion. When the bishop planned to take the bells from some of the Mojos' churches for use in the cathedral of Santa Cruz, one carayana noted that the Mojos felt "as if in each object that leaves their temples is taken, more than just an ecclesiastical property, but something sacred of their history and of their very life."[85] And when the Bolivian Congress planned to disestablish the Catholic Church in Beni by abolishing the cóngruas to its parish priests, another carayana writer felt sure the Mojos would support their priest in Trinidad. The writer credited the more than two thousand Trinitarios with conserving both the faith and civilization.[86] In both these cases, the more conservative carayanas professed admiration for the Mojos' Catholicism, and even equated it with civilization itself in much the same way as Santos Noco did.

Mojos were frequently involved with the cemetery in Trinidad, whether because of their religiosity and respect for the dead, or because the carayanas considered such work to be demeaning and unclean. In 1912, the work of the Indian grave diggers aroused the disapproval of the carayanas, who noted that they buried bodies only four palms deep and refused to step on the dirt of a grave to tamp it down. It is unclear whether this reflects a special indigenous attitude toward the dead or not.[87] In 1913, the municipal council tried to improve this situation by naming a *celador municipal* (municipal guard) with duties as caretaker of the public cemetery. The council chose the Mojo Mariano Semo for the position, again demonstrating that Mojos were considered to be the proper people to work in the cemetery.[88]

Mojos outside of Trinidad also continued to support the church. In Loreto in 1906, Corregidor Jesús Añez accused the priest, Epifanio Campos, of wrongdoing. The priest resigned and the prefect stopped paying his cóngrua. On June 15, virtually all the Mojos of Loreto met at Cacique Manuel María Muiva's house and drafted a letter to the priest of Trinidad in support of Campos, following the example of the carayana vecinos two days earlier. In contrast to the vecinos' petition, which was signed only by men, both Mojo men and women signed their letter, with men listed before women. While Mojo women may not have been exactly equal to Mojo men, in this case they participated more in Mojo society than carayana women did in theirs.[89]

One event that symbolizes a break with the Mojos' mission past is the destruction of the old Jesuit-style churches after the establishment of the Apostolic Vicariate of Beni. The new priest of San Ignacio, Estanislao V. de Marchena, had the church of that town torn down and a new one constructed in its place. He also had the Ignacianos construct a meeting house called the Belén on the church grounds, where they were to practice the ceremonies and hold the cabildo meetings that they had formerly undertaken in the cacique's house.[90] The present Romanesque church of Trinidad, which bears no resemblance to its predecessor, was finished in 1932. In preparation for the new construction, Vicar Apostolic Ramón Calvo y Marti partially demolished the old church of Trinidad in 1921. The thousands of bats which had roosted under its roof swarmed out to find new homes, an apt metaphor for the Trinitarios displaced from their residences in the center of town.[91] Yet the destruction of a building does

*Figure 6.1.* Present prefecture and Church of Trinidad. Author photograph.

not signify the end of the Mojos' culture, only that it would continue in the new homes they had made for themselves.

## Conclusion

After the Guayochería, the Mojos of the former mission towns shared a fate that was dramatically different from the Mojos of the refugee towns of San Lorenzo and San Francisco. Changing circumstances ended the earlier demand for their labor in the rubber zone, though some still plied the rivers as paddlers and pilots. The majority of Mojos saw their lives changed for the worse by a great increase in the carayana population, by a rapid growth of carayana ranching and sugar planting, and by the regulations of a mature carayana bureaucracy. The majority of Mojos became mozos, or debt peons, at the same time those in the refugee towns were asserting their freedom. Though their patrones did permit the observance of religious holidays, the mozos had little defense against abuse other than flight or killing their patrones, both of which were punished. At the same time, the carayana bureaucracy employed many different strategies to obtain unpaid labor from both mozos and free Mojos. Yet in the decades from 1890 to 1930, some Mojos took advantage of Bolivian laws to register rural properties and cattle, urban lots, and even communal properties. Many more of them participated in liberal holidays and schools while conserving their own religious festivals, cabildos, and parcialidades. The Mojos of San Ignacio enjoyed more autonomy than those of Trinidad or the other towns, and managed to preserve more of their old life and culture. Nevertheless, almost all Mojos in the former mission towns had fewer options to exercise their agency in this period than did those in the refugee towns.

On the surface, the Mojos of the old mission towns retained little visible link to their mission culture. The physical appearance of their churches and towns had changed, few Mojos owned land, and even fewer owned land communally. They seldom wore camijetas and tipoys. They were no longer the majority in their towns, and no longer controlled the region's artisanry, agriculture, cattle, or river transport. Only the cabildos and machetero dances seemed to hark back to an earlier age. Yet what is important is not the changing outward form of Mojo culture. The Mojos held on to a sense of identity created during the mission period, reflected in their continuing respect for the practices and beliefs of mission Catholicism, for cabildos, and for stable families. In the period from 1888 to 1932, the Mojos demonstrated how adaptable their culture had become. It was a culture where macheteros danced for Bolivian Independence Day, where both mozos and free Mojos practiced their religious festivals, and where native people sought, and sometimes found, ways to advance in a liberal, carayana-dominated society.

# *Conclusion*

A relatively small country in terms of population, Bolivia is nevertheless well-known as one of only two nations with a Native American majority. Until recently, it seemed that the Spanish conquest of the region had determined its social structure for all time: a creole upper class would dominate politics, economics, and culture, while Bolivia's indigenous majority would labor on haciendas, in mines, in impoverished peasant communities, and in other lower-class positions. Twentieth-century revolutionary movements offered hope for social justice but denied these people their identity by insisting that while class mattered, ethnic affiliations did not. Nothing seemed to change until indigenous people began to construct political organizations and reemphasize their ethnicity in the 1980s. Bolivia captured the world's attention by electing Evo Morales, a man who emphasized his Aymara indigenous heritage, as president in 2006. And though highland Quechua and Aymara speakers have been most visible in Bolivia's recent indigenous activism, lowland peoples have not been absent. The Mojo people participated in marches from the lowlands to La Paz in 1990 and 2011 to present their demands to the national government.

The newly invigorated indigenous activism of the Andean republics often uses the language of citizenship and rights associated with liberal discourse. In her study of Quechua-speakers in 1990s Peru, anthropologist María Elena García found a newly emerging idea of "indigenous citizenship": the notion that one can be both an equal member of a national population and a member of a culturally distinct indigenous nation.[1] Yet as this study shows, indigenous agency is nothing new, and indigenous people have creatively participated in the discourse of liberalism since the nineteenth century. Liberalism reached the Mojos in the 1840s, and their leaders began demanding equal rights *and* respect for their cultural distinctiveness by the late 1880s. Liberalism's discourse became a site of contention between whites and Indians. Bolivian white elites wanted the material progress associated with liberalism without giving up their dominant place in society. Although some creoles were initially willing to include Indian with white elites, they quickly reverted to a position that excluded Indians. Beni's white elite emerged in conjunction with the rubber boom, saw themselves as bringing progress and modernity to the region, and used liberalism to justify their assumption of control of society from an indigenous leadership they painted as backward and premodern.

Liberal reforms exposed Indians to the ideals of equality before the law, equality of opportunity, and civil rights, and the true legacy of the liberal experiment in mid-nineteenth-century Beni was the implantation of these ideas of equal rights among the Mojos. Mojo leaders used liberalism to try to gain the equality that white elites denied them. They desired the political equality promised by liberalism without giving up cabildo government or mission Catholicism, though they were willing to encourage the increased teaching of Spanish as a way to participate in national society. If any elements of the mission legacy were discarded, it was because they no longer served the Mojos' interest.

The ex-mission Indians of Beni might have stayed in their home towns if the liberal state had truly provided them with increased political, economic, and educational opportunities. When these opportunities were not forthcoming, many Beni Indians chose to migrate. The first major opportunity for migration came with the rubber boom. While traditional accounts emphasize the brutal treatment of rubber workers, and revisionist histories claim a relative lack of exploitation, this book has shown that both versions are applicable at different times and places. Before 1880, and below the Madeira-Mamoré cachuelas, rubber-gathering and transport held some attraction for the former mission Indians. Beni Indian commoners used the rubber trade to facilitate migration to a better life in Brazil, accompanying rubber patrones to escape the demands of carayana authorities and Indian cabildos. If their patrones did not treat them well, they could move to a Brazilian city, join independent Indians, or return home, and they sometimes killed their patrones as well. After 1880, and above the cachuelas, the disadvantages of participation in the rubber boom came to outweigh the benefits for Beni Indians, who were indeed harshly exploited. Taken to upriver forests rather than the Madeira, Beni

Indians had less opportunity to escape to Brazilian cities or independent groups, or to disobey or dispose of their patrones. The rubber interests also began to exercise political clout, streamlining exploitation and greatly lessening the chances for Indians to play off the national government against the local carayana elite.

In the more restrictive climate of the 1880s, many Trinitarios chose a new kind of migration. Inspired by the preaching of the ventriloquist messiah Andrés Guayocho, they set up new towns where cabildo government and mission Catholicism could be reconstructed in an environment free from carayana exploitation. Guayocho's shamanism derived from pre-Christian indigenous practices, but he spoke in the voice of Jesus Christ and re-created the important elements of mission culture. And by speaking in the voices of Frutos Nosa and other deceased Mojo leaders, Guayocho represented himself as their legitimate successor and laid a claim to their relationship with the liberal state. So despite his unorthodox approach to Christianity, he asked for a corregidor and priest for San Lorenzo, and displayed the Bolivian flag at Carnival.

This book has recounted the brutal repression of Guayocho's movement by a departmental government allied with the rubber interests. But when José Santos Noco led the reoccupation of San Lorenzo, the Trinitario refugees demonstrated that Guayocho's movement had not failed. For thirty-four years, the Lorenzanos abandoned millenarian hopes and engaged with the dominant society under Santos Noco's leadership. He and his skillfully chosen allies managed to protect Indian land and regulate labor relations, using the discourses of both liberalism and of civilization and barbarism. He also created a new, hybrid discourse by combining liberal demands for progress, educational opportunities, and instruction in the Spanish language with a continued respect for mission-style Catholicism and communal property holding. Though less successful, Mojos in Trinidad and San Ignacio also managed to protect some of their property and mission culture while struggling for equal opportunity in liberal Bolivia.

The rise and fall of Guayocho's millenarian movement may be the most compelling of the stories related in this book. The ventriloquist messiah attracted my interest to this period of Mojo history, and while putting his movement into its historical context, I discovered that it was only part of a continuum of Mojo agency in the era of liberalism and rubber. Constantly weighing labor and tax burdens against their desire to preserve their families, religion, and cabildo government, the Mojos tried to retain the best of the colonial legacy while engaging in the export economy and in liberal discourse in order to increase their opportunities in society.

Such creative adaptation allowed the Mojos to survive as a distinct ethnic group. They would face other challenges after 1932, from recruitment for the Chaco War to another boom in carayana ranching related to the ability to transport fresh meat to the Bolivian highlands by plane. The Mojos would be able to survive those trials (through migration, millenarianism, participation, and political activism) only because they had previously survived the eras of liberal reform and the rubber boom. Guayocho attracted a following because he spoke to the dreams of a people. And although the old shaman's enemies cruelly ended his life on that hot day on the Amazonian savanna in 1887, the Mojos have continued pursuing those dreams of equality and respect to this day.

# Notes

## Introduction

1. Scholars have used the concept of agency more frequently in studies of Africa and its diaspora than for Latin American indigenous peoples. My ideas of agency are informed by the introduction to Andrew Apter, *Beyond Words* (Chicago: University of Chicago Press, 2007). Important examples of earlier theoretical approaches include Eric Wolf, "Closed Corporate Peasant Communities in Mesoamerica and Central Java," *Southwestern Journal of Anthropology* 13, no. 1 (1957): 1–18, in ethnographic theory; Eric Hobsbawm, *Primitive Rebels* (New York: W. W. Norton, 1959) in Marxian theory; and James Scott, *Weapons of the Weak* (New Haven: Yale University Press, 1985) in resistance theory. Important studies of Latin American history informed by resistance theories include Steve Stern, ed., *Resistance, Rebellion, and Consciousness in the Andean Peasant World, 18th to 20th Centuries* (Madison: University of Wisconsin Press, 1987); Erick Langer, *Economic Change and Rural Resistance in Southern Bolivia, 1880–1930* (Stanford: Stanford University Press, 1989); Roger Rasnake, *Domination and Cultural Resistance* (Durham, NC: Duke University Press, 1988); John Kicza, ed., *The Indian in Latin American History* (Wilmington, DE: Scholarly Resources, 1993); William Beezley et al., eds. *Rituals of Rule, Rituals of Resistance* (Wilmington, DE: Scholarly Resources, 1994); and Catherine Legrand, "Informal Resistance on a Dominican Sugar Plantation during the Trujillo Dictatorship," *Hispanic American Historical Review* 75, no. 4 (Nov. 1995): 555–96. The limitations of resistance theory have been raised by social scientists in Deborah Poole, ed., *Unruly Order* (Boulder, CO: Westview Press, 1994); Michael Brown, "Beyond Resistance," in Anna Roosevelt, ed., *Amazonian Indians from Prehistory to the Present* (Tucson: University of Arizona Press, 1994), 287–311; Michael Brown, "On Resisting Resistance," *American Anthropologist* 98, no. 4 (Dec. 1996): 729–35; and Jocelyn Hollander and Rachel Einwohner, "Conceptualizing Resistance," *Sociological Forum* 19, no. 4 (Dec. 2004): 533–54.

2. E. Bradford Burns, *The Poverty of Progress* (Berkeley: University of California Press, 1980), represents the traditional view of the effects of liberal reforms on Indian and mestizo peasants throughout the entire region. See also the introduction to Robert Jackson, ed., *Liberals, the Church, and Indian Peasants* (Albuquerque: University of New Mexico Press, 1997) for a summary of the literature on Latin American liberalism.

3. For the great variety in European interpretations of liberalism, see Alan Bullock and Maurice Shock, eds., *The Liberal Tradition* (London: Adam and Charles Black, 1956); John Gray, *Liberalism* (Minneapolis: University of Minnesota Press, 1986); J. S. Schapiro, *Liberalism* (New York: Van Nostrand Reinhold, 1958); W. M. Simon, ed., *French Liberalism 1789–1848* (New York: Wiley, 1972); James Sheehan, *German Liberalism in the Nineteenth Century* (Chicago: University of Chicago Press, 1978); and Frederick Artz, *Reaction and Revolution, 1814–1832* (New York: Harper and Row, 1934).

4. For some Nahua leaders who became commanders in the Mexican National Guard, see Guy Thomson, "Bulwarks of Patriotic Liberalism," *Journal of Latin American Studies* 22 (1990): 31–68; "Agrarian Conflict in the Municipality of Cuetzalán (Sierra de Puebla)," *Hispanic American Historical Review* 71, no. 2 (May 1991): 205–58; and "Popular Aspects of Liberalism in Mexico, 1848–1888," *Bulletin of Latin American Research* 10, no. 3 (1991): 265–92. For additional information on participation by Mexican indigenous people, see Raymond Buve, "Political Patronage and Politics at the Village Level in Central Mexico," *Bulletin of Latin American Research* 11, no. 1 (January 1992): 1–28; and Michael Ducey, "Liberal Theory and Peasant Practice: Land and Power in Northern Veracruz, Mexico, 1826–1900," in Jackson, ed., *Liberals, the Church, and Indian Peasants*, 65–93. Andean scholars have also examined the Bolivian *caciques apoderados* (Aymara and Quechua leaders with power of attorney for their communities); see Silvia Rivera Cusicanqui, "'Pedimos la revisión de límites.' Un episodio de incomunicación de castas en el movimiento de caciques-apoderados en los Andes bolivianos, 1919–1921," in Segundo Moreno Y. and Frank Salomon, compiladores, *Reproducción y transformación de las sociedades andinas, siglos XVI–XX. Tomo II* (Quito: Ediciones ABYA-YALA, MLAL, 1991), 603–52; Leandro Condori Chura and Esteban Ticona Alejo, *El escribano de los caciques apoderados. Kasikinakan puriparunakan qillqiripa* (La Paz: Hisbol/THOA, 1992); Vitaliano Soria Choque, "Los caciques-apoderados y la lucha por la escuela (1900–1952)," in *Educación indígena: ¿ciudadanía o colonización?* (La Paz: Ediciones Ayuwiyiri, 1992), 41–78; and Laura Gotkowitz, *A Revolution for our Rights: Indigenous Struggles for Land and Justice in Bolivia, 1880–1952* (Durham, NC: Duke University Press, 2007). Brooke Larson's *Trials of Nation Making* (Cambridge: Cambridge University Press, 2004) provides an overview of creole-indigenous relations in the Liberal-era Andean republics.

5. Scholarship on frontier, ex-mission Indians includes Jane Rausch, *A Tropical Plains Frontier* (Albuquerque: University of New Mexico Press, 1984), and *The Llanos Frontier in Colombian History, 1830–1930* (Albuquerque: University of New Mexico Press, 1993); Albert Hurtado, *Indian Survival on the California Frontier* (New Haven: Yale University Press, 1988); Erick Langer and Robert Jackson, "Colonial and Republican Missions Compared," *Comparative Studies in Society and History* 30, no. 2 (April 1988): 289–303; Robert Jackson and Edward Castillo, *Indians, Franciscans, and Spanish Colonization* (Albuquerque: University of New Mexico Press, 1995); and

Cynthia Radding Murrieta, *Wandering Peoples* (Durham, NC: Duke University Press, 1997); and Erick Langer, *Expecting Pears from an Elm Tree* (Durham, NC: Duke University Press, 2009). Examples dealing specifically with the Amazon region include Jürgen Riester, *En busca de la Loma Santa* (La Paz: Editorial Los Amigos del Libro, 1976); James Jones, "Conflict between Whites and Indians on the Llanos de Moxos, Beni Department" (PhD diss., University of Florida, 1980); Zulema Lehm Ardaya, "Proyecto: 'Diagnóstico de la situación de los indígenas de Trinidad y areas cercanas'" (Trinidad: CIDDEBENI, 1987), and *Milenarismo y movimientos sociales en la Amazonia boliviana* (Santa Cruz: CIDDEBENI, APCOB, OXFAM América, 1999); Zulema Lehm Ardaya, Arnaldo Lijerón Casanovas, and Lorenzo Vare Chávez, *Diagnóstico socio-económico de los indígenas mojeños en la ciudad de Trinidad* (Trinidad: CIDDEBENI, 1990); Blanca Muratorio, *The Life and Times of Grandfather Alonso* (New Brunswick: Rutgers University Press, 1991); Fernando Santos Granero, *Etnohistoria de la alta Amazonia* (Quito: Ediciones ABYA-YALA, 1992), and *Opresión colonial y resistencia indígena en la alta Amazonia* (Quito: CEDIME, FLACSO, 1992); David Block, *Mission Culture on the Upper Amazon* (Lincoln: University of Nebraska Press, 1994); and Peter Gow, "River People," in Nicholas Thomas and Caroline Humphrey, eds., *Shamanism, History, and the State* (Ann Arbor: University of Michigan Press, 1994), 90–114.

6. Recent scholarship on Latin American frontiers includes Ana María Alonso, *Thread of Blood* (Tucson: University of Arizona Press, 1995); Donna Guy and Thomas Sheridan, eds., *Contested Ground* (Tucson: University of Arizona Press, 1998); Robert Jackson, ed., *New Views of Borderlands History* (Albuquerque: University of New Mexico Press, 1998); Gary Anderson, *The Indian Southwest, 1580–1830* (Norman: University of Oklahoma Press, 1999); James Brooks, *Captives and Cousins* (Chapel Hill: University of North Carolina Press, 2002); Andrés Reséndez, *Changing National Identities at the Frontier* (Cambridge: Cambridge University Press, 2005), and David Weber, *Bárbaros* (New Haven: Yale University Press, 2005).

## Chapter 1

1. William Denevan, *The Aboriginal Cultural Geography of the Llanos de Mojos of Bolivia* (Berkeley: University of California Press, 1966), 9–11; Daniel Santamaría, "La economía de las misiones de Moxos y Chiquitos (1675–1810)," *Ibero-Amerikanisches Archiv* 13, no. 2 (1987), 260.

2. John Hemming, *Amazon Frontier* (London: Macmillan, 1987), 277.

3. Edward Mathews, *Up the Amazon and Madeira Rivers through Bolivia and Peru* (London: S. Low, Marston, Searle and Rivington, 1879), 168; "Inundación," *El Correo del Beni*, 16 febrero 1895, 2; "1896," *El Correo del Beni*, 24 enero 1896, 1; "Inundación," *La Democracia*, 9 febrero 1907, 4; "Cuatro inundaciones en una década," *El Eco del Beni*, 31 enero 1931, 1.

4. Chris Paola, "Floods of Record," *Nature* 425 (2 Oct. 2003): 459; Rolf Aalto et al., "Episodic Sediment Accumulation on Amazonian Flood Plains Influenced by El Niño/Southern Oscillation," *Nature* 425 (2 Oct. 2003): 493–97.

5. Jones, "Conflict between Whites and Indians," 19; Allan Holmberg, *Nomads of the Long Bow* (Garden City, NY: Natural History Press, 1969), 3.

6. "Urjente rectificación del catastro," *La Democracia*, 12 julio 1913, 2–3; Lardner Gibbon, *Exploration of the Valley of the Amazon* (Washington, DC: A.O.P. Nicholson, 1854), 228–229, 253; Alcides d'Orbigny, *Descripción geográfica, histórica y estadística de Bolivia, Tomo I* (La Paz: Instituto Cultural Anglo-Boliviano, 1946), 170.

7. Jones, "Conflict between Whites and Indians," 51, n. 3.

8. Alf Hornborg, "Ethnogenesis, Regional Integration, and Ecology in Prehistoric Amazonia," *Current Anthropology* 46, no. 4 (Aug.–Oct. 2005): 589–602.

9. Julian Steward, ed., *Handbook of South American Indians*, Vol. 3 (Washington, DC: Bureau of American Ethnology, 1948), 1.

10. Holmberg, *Nomads*, 14.

11. Block, *Mission Culture*, 16–21; Alfred Métraux, "The Social Organization and Religion of the Mojo and Manasi," *Primitive Man* 16, nos. 1 and 2 (January and April 1943): 9.

12. Denevan, *Aboriginal Cultural Geography*, 40–46, 56; David Wilson, *Indigenous South Americans of the Past and Present* (Boulder, CO: Westview Press, 1999), 174; Block, *Mission Culture*, 27; Rodolfo Pinto Parada, *Pueblo de leyenda* (Trinidad: Tiempo del Bolivia, 1987), 310–11.

13. Métraux, "Social Organization and Religion," 6–7; Block, *Mission Culture*, 25–27, 86; Denevan, *Aboriginal Cultural Geography*, 59–60.

14. Métraux, "Social Organization and Religion," 10–11, 17–18; Orbigny, *Descripción*, 220. The word *chicha* was brought to the llanos by the Spanish. Local names for the beverage included *catsiamo* (Mojo), *pissisi* (Canichana), *poo-zo* (Movima), *veiqui* (Cayubaba), *uguamba* (Itonama), and *marca* (Baure). See Fray José Cardús, *Las misiones franciscanas entre los infieles de Bolivia* (Barcelona: Librería de la Inmaculada Concepción, 1886), 309, 315–18; Lehm, *Milenarismo*, 24.

15. Diego Francisco Altamirano, *Historia de la Misión de los Mójos* (La Paz: Instituto Boliviano de Cultura, Biblioteca "José Agustín Palacios," 1979), 31–33, 54–55; Francisco Eder, *Breve descripción de las reducciones de Mojos* (Cochabamba: Historia Boliviana, 1985), 110, 120; Block, *Mission Culture*, 26–27; Lehm, Lijerón, and Vare, *Diagnóstico socio-económico de los indígenas mojeños*, 11; Lehm, *Milenarismo*, 23–24, 58; Métraux, "Social Organization and Religion," 14–16; Orbigny, *Descripción*, 220; Mircea Eliade, *Shamanism* (Princeton: Princeton University Press, 1964), 23–35.

16. Eder, *Breve descripción*, 110–11; Métraux, "Social Organization and Religion," 9; Lehm, *Milenarismo*, 24; Jones, "Conflict between Whites and Indians," 159–62.

17. Wolfgang Schoop, *Ciudades bolivianas* (La Paz, Cochabamba: Editorial Los Amigos del Libro, 1981), 191–93. The settlers of Santa Cruz moved to the present Cotoca in 1604 before establishing Santa Cruz at its final site in 1621.

18. Block, *Mission Culture*, 34–38.

19. Orbigny, *Descripción*, 225.

20. Eder, *Breve descripción*, 313–17.

21. Ibid., 109–27; Lehm, *Milenarismo*, 58; Jones, "Conflict between Whites and Indians," 159–62.

22. Block, *Mission Culture*, 41, 57–58, 78, 85.

23. Ibid., 38–44, 78–85.

24. Ibid., 1–2, 125.

25. Ibid., 87–89, 94–100. In a proposal to revive the Jesuit system in the mission towns in 1905, it was suggested that the Indians be housed in multifamily houses

(*cuarteles*) according to parcialidades, indicating that this was the original arrangement in the missions. See "Correspondencia," *La Democracia*, 20 agosto 1905, 2.

26. Alberto Armani, *Ciudad de Dios y Ciudad del Sol* (México: Fondo de Cultura Económica, 1982), 103–04; Lehm, Lijerón, and Vare, *Diagnóstico socio-económico de los indígenas mojeños*, 23–25.

27. Lehm, Lijerón, and Vare, *Diagnóstico socio-económico de los indígenas mojeños*, 6.

28. Orbigny, *Descripción*, 231.

29. Block, *Mission Culture*, 67–76.

30. Orbigny, *Descripción*, 367; Alejandro Cañeque, *The King's Living Image* (New York: Routledge, 2004), 188–89.

31. Block, *Mission Culture*, 125–30, 133–41; Orbigny, *Descripción*, 236–37; Santamaría, "La economía de las misiones," 270–84.

32. Block, *Mission Culture*, 130–32, 142–48, 165–66; Santamaría, "La economía de las misiones," 258, 271; Alcides d'Orbigny, *El hombre americano* (Buenos Aires: Editorial Futuro, 1944), 345; José Luis Roca, "Mojos en los albores de la Independencia boliviana, (1810–1811)," *Historia y Cultura* 21 (1992): 187–98.

33. José Flores Moncayo, *Legislación boliviana del indio* (La Paz: Ministerio de Asuntos Campesinos, Departamento de Publicaciones del Instituto Indigenista Boliviano, 1953), 127–34; "Tarifa para los efectos y producciones de Mojos formada por la comisión nombrada al efecto por el Sõr. Presidente de este Departamento," 26 julio 1825 (BAHUAGRM Prefectura 1/2).

34. Orbigny, *Descripción*, 366; Alcides d'Orbigny, *Viaje a la América meridional . . . realizado de 1826 a 1833* (Buenos Aires: Editorial Futuro, 1945), 1311–12.

35. Orbigny, *Descripción*, 338; Actos de gobierno, correspondencia, 1841. Copiador de notas dirijidas al gobernador de la Provincia de Mojos—Año de 1841 (in Copiador de comunicaciones del año 1841), No. 72, julio 9 (BAHUAGRM Prefectura 1/29).

36. Orbigny, *Descripción*, 242–43, 380.

37. Orbigny, *El hombre americano*, 355.

38. Roca, "Mojos en los albores," 194.

39. Orbigny, *Descripción*, 245.

40. Jones, "Conflict between Whites and Indians," 100; The singular form *caray* derives from *karai*, a Guaraní term for Spaniards, which connoted a powerful, supernatural other and which originally meant "exceptional beings," "god-men," or "prestigious shamans." See Thierry Saignes, *Ava y karai* (La Paz: Hisbol, 1990), 13, 23; Manuel Limpias Saucedo, *Los gobernadores de Mojos* (La Paz: Escuela Tipográfica Salesiana, 1942), 4.

41. Block, *Mission Culture*, 60–63.

42. Franz Keller, *The Amazon and Madeira Rivers* (Philadelphia: J. P. Lippincott, 1875), 188–89.

43. Gumercindo Gómez de Arteche, *JHS misión de los P.P. Astrain, Manzanedo y Arteche, 1888*, ed. Jorge Cortés Rodríguez (Trinidad: CIDDEBENI, 1989), 16–17.

44. Arteche, *JHS misión*, 16–17; Gibbon, *Exploration*, 248; Henry Rusby, *Jungle Memories* (New York: Whittlesey House, 1933), 174.

45. Schoop, *Ciudades bolivianas*, 223.

46. Orbigny, *Descripción*, 246; Gibbon, *Exploration*, 230.

47. Arteche, *JHS misión*, 16.

48. Enrique Jordá Arias, "La Semana Santa y la cruz en los pueblos Mojos del Beni (Bolivia)," in Frans Damen and Esteban Judd Zanon, eds., *Cristo crucificado en los pueblos de América Latina* (Cusco-Sicuani: Ed. Instituto de Pastoral Andina; Quito: Ed. ABYA-YALA, 1992), 205.

49. Alcides Parejas Moreno, "Las misiones jesuíticas de Chiquitos," in *Los Bolivianos en el tiempo* (Sucre: Universidad Andina Simón Bolívar, Instituto de Estudios Andinos y Amazónicos, 1993), 132; Corregidor José María Fresco, Inventario de bienes de la iglesia de San Pedro (ANB MI 1844 T. 101 No. 23, f. 85v.–86); Arteche, *JHS misión*, 15; Gibbon, *Exploration*, 230.

50. Orbigny, *Descripción*, 322, 369.

51. Gibbon, *Exploration*, 247–48.

52. Rogers Becerra Casanovas, *Reliquias de Moxos* (La Paz: Empresa Editora "Proinsa," 1978), 9–12; Antonio Paredes-Candia, *La danza folklórica en Bolivia* (La Paz: Libreria-Editorial "Popular," 1991), 35–36.

53. Becerra, *Reliquias de Moxos*, 13–15, 79–87.

54. Ibid., 41–45.

55. Gibbon, *Exploration*, 242–43.

56. Orbigny, *Descripción*, 369–73, 246–47.

57. Ibid., 379; Francisco José de Lacerda e Almeida, "Memoria a respeito dos Rios Baures, Branco, da Conceição, de S. Joaquim, Itonamas e Maxupo; e das três missões da Magdalena, da Conceição e de S. Joaquim," *Revista do Instituto Histórico e Geográfico Brasileiro* 12 (1849): 115. Some cuarteles had tile roofs by the end of the mission period; see Eder, *Breve descripción*, 356.

58. "Relación del numero de talleres, y artesanos que existen en el pueblo de Trinidad y demás poblaciones del Nuevo Dpto. del Beni" (ANB MI 1843 T.96 N.48).

59. Block, *Mission Culture*, 57, 155; Santamaría, "La economía de las misiones," 161; Lehm, "Proyecto," 208–11.

60. Jones, "Conflict between Whites and Indians," 76, 179–88; Denevan, *Aboriginal Cultural Geography*, 32, 96; Gibbon, *Exploration*, 261; Orbigny, *Descripción*, 379.

61. José de Borja, "Razón detallada de las obras públicas realizadas en las distintas provincias del Beni 1845," 7 enero 1845 (ANB MI T.107 N.33).

62. Tim Ingold, *Hunters, Pastoralists, and Ranchers* (Cambridge: Cambridge University Press, 1980), 2–5.

63. Terry Jordan, *North American Cattle Ranching Frontiers* (Albuquerque: University of New Mexico Press, 1993), 23–35, 93–96, 107, 137, 142; Block, *Mission Culture*, 57–58, 93, 99, 136, 156; Gibbon, *Exploration*, 244–45, 252; Mathews, *Up the Amazon and Madeira Rivers*, 151–52.

64. Orbigny, *Descripción*, 367–368; Orbigny, *Viaje*, 1311.

65. Limpias, *Los gobernadores de Mojos*, 25.

## Chapter 2

1. Langer, *Economic Change*, 14–21; Robert Jackson, *Regional Markets and Agrarian Transformation in Bolivia* (Albuquerque: University of New Mexico Press, 1994), 55–58, 70–89; Erick Langer and Robert Jackson, "Liberalism and the Land Question in Bolivia, 1825–1920," in Jackson, ed. *Liberals, the Church, and Indian Peasants*, 171–72, 180–85.

2. Limpias, *Los gobernadores de Mojos*, 4; Langer, *Economic Change*, 15; Manuel Carrasco, *José Ballivián, 1805–1852* (Buenos Aires: Machette, 1960), 182–84; Janet Groff Greever, *José Ballivián y el oriente boliviano* (La Paz: Empresa Editora Siglo, 1987), 40.

3. Orbigny, *Viaje*, 1312, 1318, 1451–52.

4. Limpias, *Los gobernadores de Mojos*, 4–5.

5. Laura Gotkowitz, "Within the Boundaries of Equality" (PhD diss., University of Chicago, 1998), 64–65; Rossana Barragán, "Ciudadanía y elecciones, convenciones y debates," in Rossana Barragán and José Luis Roca, eds., *Regiones y poder constituyente en Bolivia* (La Paz: Informe sobre el Desarrollo Humano en Bolivia, 2005), 289–90, 296–97.

6. Limpias, *Los gobernadores de Mojos*, 24–26, 65.

7. Manuel de la Cruz Mobo to Juez de Letras of Trinidad, 29 noviembre 1856 (ANB MI T.158 N.17); Carmelo M. del Rivero to Ministro del Interior y Culto, 4 diciembre 1856 (ANB MI T.158 N.17).

8. Lehm, Lijerón, and Vare, *Diagnóstico socio-económico de los indígenas mojeños*, 6.

9. Limpias, *Los gobernadores de Mojos*, 47–48, 59–62, 75–93.

10. Ibid., 38, 48, 176–77; Gibbon, *Exploration*, 243–46; *Padrón jeneral de indijenas contribuyentes de la Provincia de Mojos, Año de 1854* (ANB RV Beni 13, T.1) f.200v.; Revisita of Beni, 1861 (ANB TNC RV N.10 1861), f.206v., 248.

11. Limpias, *Los gobernadores de Mojos*, 59–62, 120–22.

12. Ibid. 79–93, 127; Gibbon, *Exploration*, 246.

13. "Legajo de autos de visita en el Dpmto. del Beni, por el Yltmo. Sr. Obispo D.D. Manuel Angel del Prado, Año 1850," ABHASC.

14. Limpias, *Los gobernadores de Mojos*, 4, 49–51, 56; Flores, *Legislación boliviana*, 169.

15. Limpias, *Los gobernadores de Mojos*, 51; Gregorio José de Vásques, "Resumen general," 30 junio 1845 (ANB MI T.107 N.33); *Matrícula jeneral del Beni para el Tribunal de Cuentas* (ANB RV Beni 1874 3), 45; Gibbon, *Exploration*, 232–33.

16. Limpias, *Los gobernadores de Mojos*, 139–40; Jones, "Conflict between Whites and Indians," 99–101; Antonio Carvalho Urey, *Beni, ensayo de interpretación histórica* (La Paz: Imprenta y Librería "Renovación" Ltda., 1983), 32–33; Block, *Mission Culture*, 155.

17. Manuel Ponferrada, "Estado que manifiesta la quiebra que han padecido los valores de los productos de la Provincia de Mojos en los dos años de 1837 y 1838, comparados con los anteriores de 1835 y 1836" (ANB MI T.115 N.37).

18. Actos de gobierno, correspondencia, 1841. Copiador de notas dirijidas al gobernador de la Provincia de Mojos—Año de 1841 (in Copiador de comunicaciones del año 1841), No. 38, Abril 29 (BAHUAGRM Prefectura 1/29).

19. José de Borja to Ministro del Interior, 7 octubre 1843 (ANB MI T.96 N.48); Limpias, *Los gobernadores de Mojos*, 4, 114; José de Borja to Ministro del Interior, 3 junio 1844 (ANB MI T.101 N.23); José de Borja to Ministro del Interior, 3 abril 1845 (ANB MI T.107 N.33).

20. Limpias, *Los gobernadores de Mojos*, 64, 73, 115, 134; Gibbon, *Exploration*, 233, 245.

21. "Orden y método que deberá observarse en los trabajos de sembradíos cultivo y cosecha que se hacen en los Cantones de esta provincia de cuenta del Estado" (ANB MI T.101 N.23).

22. Flores *Legislación boliviana*, 127–34.

23. Jones, "Conflict between Whites and Indians," 102–3; Flores, *Legislación boliviana*, 169; Limpias, *Los gobernadores de Mojos*, 134–36; José Manuel Gutiérrez, *Colección oficial del año de 1858* (Cochabamba: Tipografía de Quevedo, 1859), 179–80.

24. Limpias, *Los gobernadores de Mojos*, 135, 148, 150; Lehm, Lijerón, and Vare, *Diagnóstico socio-económico de los indígenas mojeños*, 10.

25. Jones, "Conflict between Whites and Indians," 103–4.

26. José de Borja, "Disposiciones que deberían rejir en la salida y conducion de los correos ordinarios que se despachen de este Departamento a los del interior de la Republica por las vias del Chapare y el Piray," 23 noviembre 1843 (ANB MI T.96 N.48).

27. Limpias, *Los gobernadores de Mojos*, 50–51, 64.

28. Gibbon, *Exploration*, 227.

29. Limpias, *Los gobernadores de Mojos*, 151–54.

30. Block, *Mission Culture*, 157; Jones, "Conflict between Whites and Indians," 105–6; Franz Keller, *Amazon and Madeira Rivers*, 182–85.

31. Gibbon, *Exploration*, 246.

32. Limpias, *Los gobernadores de Mojos*, 31, 44, 111, 114.

33. José de Borja to Ministro del Interior, 10 agosto 1843 (ANB MI T. 96, No. 48); Gibbon, *Exploration*, 179, 197, 241–42; José de Borja to Ministro del Interior, 9 octubre 1843 (ANB MI T.96, N.48); Limpias, *Los gobernadores de Mojos*, 25–26.

34. Ross Jamieson, *Domestic Architecture and Power* (New York: Kluwer Academic/Plenum Publishers, 2000), 49–52, 97–98.

35. José de Borja, "Razón detallada de las obras públicas realizadas en las distintas provincias del Beni 1845," 7 enero 1845 (ANB MI T.107 N.33).

36. Ibid.

37. Ibid.; Gibbon, *Exploration*, 237–238.

38. Limpias, *Los gobernadores de Mojos*, 33, 111; Nicolás Rojas, "Razón de las obras públicas que se trabajaron en el Departm[en]to desde 1° de Julio h[as]ta 31 de Diciembre del año 52" and "Razón de las obras públicas q[u]e se han trabajado en el Departm[en]to desde el 1° de En[er]o del p[resen]te año, h[as]ta fin del presente mes" (ANB MI T.146 N.19); Gibbon, *Exploration*, 229; Schoop, *Ciudades bolivianas*, 223–224.

39. Antonio Carvalho Urey, *Visión del Beni* (Cochabamba: Editorial Serrano Ltda., 1978), 28; Rojas, "Razón . . . desde 1° de Julio h[as]ta 31 de Diciembre del año 52" and "Razón . . . desde el 1° de En[er]o del p[resen]te año, h[as]ta fin del presente mes" (ANB MI T.146 N.19); Franz Keller, *Amazon and Madeira Rivers*, 187, 80–82; Mathews, *Up the Amazon and Madeira Rivers*, 123, 127.

40. José de Borja, "Razón detallada de las obras públicas realizadas en las distintas provincias del Beni 1845," 7 enero 1845 (ANB MI T.107 N.33); Limpias, *Los gobernadores de Mojos*, 37–38, 111, 127.

41. Block, *Mission Culture*, 165; Oficina Nacional de Inmigración, Estadística y Propaganda Geográfica, *Censo general de la población de la República de Bolivia según el empadronamiento de 1° de septiembre de 1900*, Vol. 1 (La Paz: Taller Tipo-Litográfico de José M. Gamarra, 1902) 2; Gibbon, *Exploration*, 232–33; Lehm, "Proyecto," 208.

42. Rosa Boccolini and Rosario Jiménez, *Estadísticas de contribución indígena en Bolivia, 1770–1902*, 2 (Lima: Biblioteca Andina, 1979), 248.
43. Orbigny, *Descripción*, 246, 251–52, 254.
44. "Legajo de autos de visita en el Dpmto. del Beni, por el Yltmo. Sr. Obispo D.D. Manuel Angel del Prado, Año 1850," f.3v–9; Schoop, *Ciudades bolivianas*, 222; Gibbon, *Exploration*, 242.
45. Limpias, *Los gobernadores de Mojos*, 4, 51; Block, *Mission Culture*, 154.
46. Gibbon, *Exploration*, 234; Block, *Mission Culture*, 154.
47. José de Borja, "Relación ecsajta que remite el Prefecto del Departamento del Beni al Ministro del Interior en cumplimiento de la orden circular de 4 de Agosto ultimo marcada con el N 18," 2 noviembre 1843 (ANB MI T.96 N.48); Investigation of Cecilio Moreno (ANB MI T.101 N.23 f.15–18); Manuel de la Cruz Mobo to Juez de Letras of Trinidad, 29 noviembre 1856 (ANB MI T.158 N.17); Revisita of Beni, 1861 (ANB TNC RV No. 10 1861), f. 237v.
48. José de Borja to Ministro del Interior, 27 abril 1845; Rafael de la Borda to Ministro del Interior, 3 mayo 1845; José Agustín Palacios to Prefect of Beni, 20 septiembre 1845; Rafael de la Borda to Ministro del Interior, 30 septiembre 1845 (all ANB MI T.107 N.33); Limpias, *Los gobernadores de Mojos*, 48.
49. Manuel de la Cruz Mobo to Juez de Letras of Trinidad, 29 noviembre 1856 (ANB MI T.158 N.17).
50. Barragán, "Ciudadanía y elecciones," 285–86, 298–300, 305–07.
51. Gibbon, *Exploration*, 246, 253–54.
52. "Legajo de autos de visita en el Dpmto. del Beni, por el Yltmo. Sr. Obispo D.D. Manuel Angel del Prado, Año 1850."
53. Rafael de la Borda to Ministro del Interior, 16 junio 1846 (ANB MI 1846 T.115 N.37).
54. Limpias, *Los gobernadores de Mojos*, 101.
55. Actos de gobierno, correspondencia, 1841. Copiador de notas dirijidas al gobernador de la Provincia de Mojos—Año de 1841 (in Copiador de comunicaciones del año 1841), No. 76, Julio 9 (BAHUAGRM Prefectura 1/29); Carvalho, *Visión del Beni*, 28; Limpias, *Los gobernadores de Mojos*, 39–40, 70; "Legajo de autos de visita en el Dpmto. del Beni, por el Yltmo. Sr. Obispo D.D. Manuel Angel del Prado, Año 1850"; Mariano Arroyo to Prefect of Beni, 27 febrero 1844 (ANB MI T.101 N.23); Gibbon, *Exploration*, 236; Franz Keller, *Amazon and Madeira Rivers*, 187, 80–82.
56. Mariano Arroyo to Prefect of Beni, 27 febrero 1844 (ANB MI T.101 N.23); Limpias, *Los gobernadores de Mojos*, 40–41, 56–57.
57. Limpias, *Los gobernadores de Mojos*, 64–65.
58. Gibbon, *Exploration*, 243–45.
59. Ibid., 245–46.
60. Limpias, *Los gobernadores de Mojos*, 134.
61. Carvalho, *Beni*, 33; Juan Coimbra, *Siringa* (Santa Cruz: Librería Editorial América, 1989), 51–52.
62. Limpias, *Los gobernadores de Mojos*, 156–58.
63. Lehm, Lijerón, and Vare, *Diagnóstico socio-económico de los indígenas mojeños*, 15; Block, *Mission Culture*, 169; República Boliviana, Departamento del Beni, *Padron jeneral de indijenas contribuyentes de la Provincia de Mojos, Año de 1854* (ANB RV Beni 13, T. 1).

64. Revisita of Beni, 1861 (ANB TNC RV No. 10 1861); *Matrícula jeneral del Beni para el Tribunal de Cuentas* (ANB RV Beni 1874 3).

65. Limpias, *Los gobernadores de Mojos*, 50, 56, 154–58.

66. In the environs of Trinidad, the Carpinteros and Siyaboconos held communal land in 1845, as did the Sacristanes in 1887 and the Tejedores in 1926. See Limpias, *Los gobernadores de Mojos*, 38; Miguel Antelo, "Sucesos del Beni," *Los Debates*, 10 noviembre 1887, 3; and "Prevención," *El Eco del Beni*, 27 noviembre 1926, 4.

67. Revisita of Beni, 1861 (ANB TNC RV No.10 1861); "Legajo de autos de visita en el Dpmto. del Beni, por el Yltmo. Sr. Obispo D.D. Manuel Angel del Prado, Año 1850;" Flores, *Legislación boliviana*, 70.

68. Revisita of Beni, 1861 (ANB TNC RV N.10 1861).

69. Limpias, *Los gobernadores de Mojos*, 16.

70. Ibid., 96; Revisita of Beni, 1861 (ANB TNC RV N.10 1861) f. 81–89; José de Borja, "Razón de los Chimanes bautisados principiando desde el 20 de Agosto de 1843 hasta el 19 de Diciembre del mismo año a ecepcion de las mugeres que fueron bautizadas en 4 de Noviembre y 17 de Diciembre," 1 junio 1844 (ANB MI T.101 N.23) f. 8–9v; Ladislao Marín to Prefect of Beni, 16 diciembre 1845 (MI T.107 N.33).

71. José de Borja, "Razón detallada de las obras públicas realizadas en las distintas provincias del Beni 1845," 7 enero 1845 (ANB MI T.107 N.33); José Antonio Hurtado, "Padrón o senso correspondiente al primer semestre del año de 1845 que manifieste el numero total de almas existentes en la Colonia de este P[ue]blo con especificación de nombres, edades, oficios, estados, y sexos," 1 agosto 1845 (ANB MI T.107 N.33).

72. Limpias, *Los gobernadores de Mojos*, 21–23, 30; José de Borja to Ministro del Interior, 27 noviembre 1843 (ANB MI T.96, N.48).

73. Rafael de la Borda to Ministro del Interior, 2 agosto 1845 (ANB MI T.115 N.37); Rafael de la Borda to Ministro del Interior, 2 enero 1846 (ANB MI T.115 N.37).

74. Limpias, *Los gobernadores de Mojos*, 25; José de Borja to Ministro del Interior, 1 junio 1844 (ANB MI T.101 N.23); Pablo Camino, "Diario que lleba el que subscribe en la empresa del descubrim[ien]to del camino por tierra de este cantón al rio grande frontera del Departamento de Santa Cruz," 31 diciembre 1845 (ANB MI T.107 N.33a); Rafael de la Borda to Ministro del Interior, 2 enero 1846 (ANB MI T.115 N.37).

75. Limpias, *Los gobernadores de Mojos*, 65.

76. José de Borja to Ministro del Interior, 3 abril 1845 (ANB MI T.107, N.33); José María Fresco to Prefect of Beni, 29 marzo 1845 (ANB MI T.107, N.33); Actos de gobierno, correspondencia, 1841. Copiador de notas dirijidas al gobernador de la Provincia de Mojos—Año de 1841 (in Copiador de comunicaciones del año 1841), No. 49, Mayo 31 (BAHUAGRM Prefectura 1/29); José de Borja to Ministro del Interior, 23 diciembre 1844 (ANB MI T.101, N.23) f. 112; León Hurtado to Prefect of Beni, 20 diciembre 1845 (ANB MI T.107 N.33a).

77. José Guillermo Campos to Juez de Paz of Exaltación, 6 octubre 1846 (ANB MI T.115 N.37); José Guillermo Campos to Juez de Paz of Exaltación, 19 noviembre 1846 (ANB MI T.115 N.37); José Guillermo Campos to Prefect of Beni, 30 noviembre 1846 (ANB MI T.115 N.37).

78. Limpias, *Los gobernadores de Mojos*, 39, 42, 52.

79. León Hurtado to Prefect of Beni, 20 diciembre 1845 (ANB MI T.107 N.33a); Actos de gobierno, correspondencia, 1841. Copiador de notas dirijidas al gobernador

de la Provincia de Mojos—Año de 1841 (in Copiador de comunicaciones del año 1841), No. 49, Mayo 31 (BAHUAGRM Prefectura 1/29); José de Borja to Ministro del Interior, 3 abril 1845 (ANB MI T.107, N.33); José María Fresco to Prefect of Beni, 29 marzo 1845 (ANB MI T.107, N.33).

80. Limpias, *Los gobernadores de Mojos*, 72; Gibbon, *Exploration*, 258–60.

81. Gibbon, *Exploration*, 224, 231, 234.

## Chapter 3

1. Hemming, *Amazon Frontier*, 271–73; Michael Stanfield, *Red Rubber, Bleeding Trees* (Albuquerque: University of New Mexico Press, 1998), 19–20; Warren Dean, *Brazil and the Struggle for Rubber* (Cambridge: Cambridge University Press, 1987), 4, 9–10; Daniel Suárez, *Manifiesto del ex-prefecto del Beni ante la opinión pública* (Trinidad: Imprenta Suárez y Hnos., 1887), 6.

2. Herbert Klein, *Bolivia* (New York: Oxford University Press, 1992), 131; Francisco José Furtado, *Relatorio que a'assemblea legislativa provincial do Amazonas apresentou na abertura da sessão ordinaria em o dia 3 de maio de 1859* (Manaus: Typographia de Francisco José da Silva Ramos, 1859), 5; Adolfo de Barros Cavalcanti de Albuquerque Lacerda, *Relatorio apresentado a'assemblea legislativa da provincia do Amazonas na sessão ordinario do 1º de outubro de 1864* (Pernambuco: Typographia de Manoel Figueiroa de Faria and Filho, 1864), 44; Manoel Clementino Carneiro da Cunha, *Relatorio apresentado á Assemblea Legislativa da provincia do Amazonas pelo Exmº Senr. Dr. Manoel Clementino Carneiro da Cunha, presidente da mesma provincia, na sessão ordinaria de 3 de maio de 1862* (Pará: Typographia de Frederico Carlos Rhossard, 1862), Documento N.3, 3; Adolpho de Barros Cavalcanti de Albuquerque Lacerda, *Relatorio com que o illustrissimo e excellentissimo senhor Dr. Adolpho de Barros Cavalcanti de A. Lacerda entregou a administração da provincia do Amazonas ao illm. e exm. sr. Tenente Coronel Innocencio Eustaquio Ferreira de Araujo* (Recife: Typ. do Jornal do Recife, 1865), 17, 23; João Wilkens de Mattos, *Relatorio com que o Exmº Sr. Presidente da provincial do Amazonas, Teniente-Coronel João Wilkens de Mattos, abrio a Assambléa Legislativa Provincial no dia 4 de abril de 1869* (Manaus: Typographia do Amazonas de Antonio da Cunha Mendes, 1869), 47.

3. Cônego Francisco Bernardino de Souza, *Lembranças e curiosidades do valle do Amazonas* (Pará: Typografía do Futuro, 1873), 247, 64; Aureliano Cândido Tavares Bastos, *O vale do Amazonas* (São Paulo: Companhia Editora Nacional, 1975), 130–31, 164; Barros Cavalcanti de Albuquerque Lacerda, *Relatorio . . . de 1864*, 41.

4. Louis and Elizabeth Cabot Cary Agassiz, *A Journey in Brazil* (Boston: Tickner and Fields, 1868), 178–79, 303–06; James Orton, *The Andes and the Amazon* (New York: Harper and Brothers, 1870), 247.

5. María del Pilar Gamarra Tellez, "Haciendas y peones en el régimen hacendatario gomero boliviano" (Tesis de Licenciatura, Universidad Mayor de San Andrés, La Paz, 1992), 36–38; Austin Coates, *The Commerce in Rubber* (Singapore: Oxford University Press, 1987), 48.

6. J. Valerie Fifer, "The Empire Builders," *Journal of Latin American Studies* 2, no. 2 (1970): 117.

7. Franz Keller, *Amazon and Madeira Rivers*, 47, 200n.; Fray Jesualdo Maccheti, *Diario del viaje fluvial del Padre Fray Jesualdo Maccheti* (La Paz: Imprenta de "El Siglo Industrial," n.d.), 52–64.

8. Mathews, *Up the Amazon and Madeira Rivers*, 19, 85, 94, 134; Charles Brown and William Lidstone, *Fifteen Thousand Miles on the Amazon and its Tributaries* (London: Stanford, 1878), 333, 338, 348–50, 353; Dean, *Brazil*, 16; W. L. Schurz, "The Distribution of Population in the Amazon Valley," *Geographical Review* 15, no. 2 (April 1925): 222; Francisco de Souza Mesquita, "Thesouraria da commissão encarregada de agenciar donativos para as obras do Hospital de Caridade em Manáos, 2 de março de 1875," in Domingos Monteiro Peixoto, *Relatorio com que o Exm. Sr. Dr. Domingos Monteiro Peixoto entregou a Administração da Provincia ao Exm. Sr. 1° vice-presidente capitão de mar e guerra Nuno Alves Pereira de Mello Cardoso em 16 de março do 1875* (Manaus: Typ. do Commercio do Amasonas, 1875).

9. João Severino da Fonseca, *Voyage autour du Brésil* (Rio de Janeiro: Librairie A. Lavignasse Filho, 1899), 279–82, 287–88.

10. Neville Craig, *Recollections of an Ill-Fated Expedition to the Headwaters of the Madeira River in Brazil* (Philadelphia: J. B. Lippincott, 1907), 130–32, 248–50, 252, 256, 261–62, 291.

11. Júlio Pinkas, *Relatorio apresentado a s. ex. o Sr. Conselheiro João Ferreira de Moura* (Rio de Janeiro: Imprensa Nacional, 1885), 32–33, 108–09, Annexo N.3, 17; Comissão de Estudos da Estrada de ferro do Madeira e Mamoré, *Do Rio de Janeiro ao Amazonas e Alto Madeira itinerario e trabalhos da comissão de estudos da Estrada de Ferro do Madeira e Mamoré* (Rio de Janeiro: Typ. a Vap. de Soares and Niemeyer, 1885), 80–84.

12. Coimbra, *Siringa*, 8; Fray Nicolás Armentia, *Diario de sus viages a las tribus comprendidas entre el Beni y Madre de Dios y en el Arroyo de Ivon en los años de 1881 y 1882* (La Paz: "Tipografía Religiosa," 1883), 5–14; Limpias, *Los gobernadores de Mojos*, 184–85, 191–92, 208, 223–24; Fifer, "Empire Builders," 118–22.

13. Fifer, "Empire Builders," 119–24.

14. Limpias, *Los gobernadores de Mojos*, 226–31; Fifer, "Empire Builders," 125–26; Pinkas, *Relatorio apresentado*, Annexo N.3, 18–19.

15. Fifer, "Empire Builders," 126–29; Limpias, *Los gobernadores de Mojos*, 203.

16. Schurz, "Distribution of Population," 217–18.

17. Michael Taussig, *Shamanism, Colonialism, and the Wild Man* (Chicago: University of Chicago Press, 1987); Stanfield, *Red Rubber, Bleeding Trees*.

18. Barbara Weinstein, *The Amazon Rubber Boom* (Stanford: Stanford University Press, 1983), 25–26, 189.

19. For primary sources, see Arteche, *JHS misión*, 17; José and Francisco Keller, *Exploración del Río Madera en la parte comprendida entre la Cachuela de San Antonio y la embocadura del Mamorè, por los injenieros brasileros José y Francisco Keller* (La Paz: Imprenta de la Union Americana—Por Cèsar Sevilla, 1870), 20–21; and Mathews, *Up the Amazon and Madeira Rivers*, 131–32, 160. For secondary sources see Rodolfo Pinto Parada, *Rumbo al Beni* (Cochabamba: Editorial Serrano, 1978), 39; and Fifer, "Empire Builders," 130.

20. E. Bradford Burns, *A History of Brazil* (New York: Columbia University Press, 1980), 331; Maccheti, *Diario del viaje fluvial*, 59; José de Miranda da Silva Reis, *Relatorio apresentado a Assembléa Legislativa Provincial do Amasonas na primeira sessão*

*da 11ª legislatura no dia 25 de março de 1872 pelo presidente da provincia o Exmº Sr. Dr. José de Miranda da Silva Reis* (Manaus: Typographia de Gregorio José de Moraes, 1872), 36; Domingos Monteiro Peixoto, *Falla diririgida [sic] á Assembléa Legislativa Provincial do Amasonas na segunda sessão da 11ª Legislatura em 25 de março de 1873 pelo presidente da provincia, bacharel Domingos Monteiro Peixoto* (Manaus: Typ. do Commercio do Amazonas de Gregorio Joze de Moraes, 1873), 40–41; Domingos Jacy Monteiro, *Relatorio apresentado ao exmº Sr. Dr. Agesiláo Pereira da Silva, presidente da provincia do Amazonas pelo Dr. Domingos Jacy Monteiro, depois de ter entregue a admimistração [sic] da provincia em 26 de maio de 1877* (Manaus: Typ. do "Amazonas" de José Carneiro dos Santos, 1878), 35; Limpias, *Los gobernadores de Mojos*, 296.

21. Franz Keller, *Amazon and Madeira Rivers*, 68, 128; Maccheti, *Diario del viaje fluvial*, 3; Mathews, *Up the Amazon and Madeira Rivers*, 14, 16, 152, 161–62; José and Francisco Keller, *Exploración del Río Madera*, 56.

22. Franz Keller, *Amazon and Madeira Rivers*, 41; Maccheti, *Diario del viaje fluvial*, 3; Mathews, *Up the Amazon and Madeira Rivers*, 32–34; Craig, *Recollections of an Ill-Fated Expedition*, 261, 316; George Church, *Aborigines of South America* (London: Chapman and Hall, 1912), 109; Fifer, "Empire Builders," 129 n. 23.

23. Franz Keller, *Amazon and Madeira Rivers*, 41–43; Maccheti, *Diario del viaje fluvial*, 3, 12, 19, 24; Mathews, *Up the Amazon and Madeira Rivers*, 48; Church, *Aborigines of South America*, 109; Fonseca, *Voyage autour du Brésil*, 250–51.

24. Craig, *Recollections of an Ill-Fated Expedition*, 287–88, 316–17, 323; Franz Keller, *Amazon and Madeira Rivers*, 55–70; Mathews, *Up the Amazon and Madeira Rivers*, 33.

25. Franz Keller, *Amazon and Madeira Rivers*, 86–91; Mathews, *Up the Amazon and Madeira Rivers*, 95–105; Church, *Aborigines of South America*, 109–10.

26. Franz Keller, *Amazon and Madeira Rivers*, 83, 86, 147; Maccheti, *Diario del viaje fluvial*, 6–8, 19–21, 26, 30, 32, 43, 46; Mathews, *Up the Amazon and Madeira Rivers*, 36, 50, 61, 73, 96–97, 106, 108.

27. Gibbon, *Exploration*, 76, 303–05; Franz Keller, *Amazon and Madeira Rivers*, 50–52; Mathews, *Up the Amazon and Madeira Rivers*, 21; Maccheti, *Diario del viaje fluvial*, 6, 49, 52–53.

28. Mathews, *Up the Amazon and Madeira Rivers*, 78–79, 116; José and Francisco Keller, *Exploración del Río Madera*, 5.

29. Franz Keller, *Amazon and Madeira Rivers*, 59; Maccheti, *Diario del viaje fluvial*, 44; Barros Cavalcanti de Albuquerque Lacerda, *Relatorio com que o illustrissimo e excellentissimo senhor Dr. Adolpho de Barros Cavalcanti de A. Lacerda entregou a administração da provincia do Amazonas*, 21; Mathews, *Up the Amazon and Madeira Rivers*, 45, 116, 136; Church, *Aborigines of South America*, xvii, 200; Craig, *Recollections of an Ill-Fated Expedition*, 322; Fifer, "Empire Builders," 121, 130.

30. Maccheti, *Diario del viaje fluvial*, 11, 17–18, 22–23; Craig, *Recollections of an Ill-Fated Expedition*, 132–33, 317, 323, 325–29, 332, 342, 355–56, 359, 371–72, 377, 399, 405, 408, 473–74; Franz Keller, *Amazon and Madeira Rivers*, 48, 91–92; Pinkas, *Relatorio apresentado*, Annexo N.7, 40, Annexo N.8, 46; José and Francisco Keller, *Exploración del Río Madera*, 11; Mathews, *Up the Amazon and Madeira Rivers*, 51–54; Fonseca, *Voyage autour du Brésil*, 270.

31. Maccheti, *Diario del viaje fluvial*, 17; Mathews, *Up the Amazon and Madeira Rivers*, 34–35, 46, 105; José and Francisco Keller, *Exploración del Río Madera*, 15, 56;

Manuel Macedonio Salinas, *Navegación de los ríos de Bolivia confluentes del Madera y Amazonas y colonización* (Cochabamba: Imprenta de Gutiérrez, 1871), 17; Rusby, *Jungle Memories*, 326–28.

32. Craig, *Recollections of an Ill-Fated Expedition*, 250, 258, 262; Pinkas, *Relatorio apresentado*, 109.

33. Craig, *Recollections of an Ill-Fated Expedition*, 130, 357–58; Mathews, *Up the Amazon and Madeira Rivers*, 15, 24–25; Franz Keller, *Amazon and Madeira Rivers*, 118; Maccheti, *Diario del viaje fluvial*, 54.

34. Mathews, *Up the Amazon and Madeira Rivers*, 54, 85, 133; Craig, *Recollections of an Ill-Fated Expedition*, 130–31, 345, 356–57, 373–76; Pinkas, *Relatorio apresentado*, 109, Annexo N.3, 19; Maccheti, *Diario del viaje fluvial*, 2, 57.

35. Limpias, *Los gobernadores de Mojos*, 215; Pinkas, *Relatorio apresentado*, 111.

36. Craig, *Recollections of an Ill-Fated Expedition*, 345, 356–57; Pinkas, *Relatorio apresentado*, 110–11; Franz Keller, *Amazon and Madeira Rivers*, 52, 42; Arteche, *JHS misión*, 17.

37. Mathews, *Up the Amazon and Madeira Rivers*, 14–15; Ladislao Cabrera, *Navegación fluvial de Trinidad a Villa Bella* (Santa Cruz: La Estrella del Oriente, 1889), 22; Dwight Heath, "Camba" (PhD diss., Yale University, 1959), 120–23; Rusby, *Jungle Memories*, 249–50, 309.

38. Franz Keller, *Amazon and Madeira Rivers*, 47–49, 118.

39. Barros Cavalcanti de Albuquerque Lacerda, *Relatorio . . . de 1864*, 44. For an example of Frisch's photography, see Erika Billeter, *Canto a la realidad* (Barcelona: Lundwerg; Madrid: Casa de América, 1993), 70.

40. Franz Keller, *Amazon and Madeira Rivers*, 42.

41. Maccheti, *Diario del viaje fluvial*, 72; João Wilkens de Mattos, *Relatorio lido pelo Exm^o^ Sr. Presidente da Provincia do Amazonas, tenente-coronel João Wilkens de Mattos, na sessão d'abertura da Assembléa Legislativa Provincial á 25 de março de 1870* (Manaus: Typographia do Amazonas de Antonio da Cunha Mendes, 1870), 33; Luiz Martins da Silva Coutinho, *Relatorio apresentado ao Illm. e Exm. Sr. Presidente da provincial do Amasonas, Dr. José de Miranda da Silva Reis pelo Director das Obras Publicas da mesma provincial, Luiz Martins da Silva Coutinho* (Manaus: Typographia de Gregorio José de Moraes, 1871).

42. Feliciano Antonio Benjamin, *Relatorio apresentado ao exm. sr. Dr. Presidente da Provincia Domingos Monteiro Peixoto, pelo Engenheiro Bacharel Feliciano Antonio Benjamin, membro da Commissão de fiscalisação da estrada de ferro do Madeira e Mamoré e de medição e demarcação de terras no Rio Madeira, actualmente encarregado da mesma commissão* (Manaus: Typographia do Commercio do Amasonas, 1874), 4–5; Barão de Maracajú, *Falla com que abrio no dia 25 de agosto de 1878 a 1ª sessão da 14ª legislatura da Assembléa Legislativa Provincial do Amazonas o exmº sr. Barão de Maracajú, presidente da provincia* (Manaus: Typ. do Amazonas, 1878), 57; Barão de Maracajú, *Falla com que o exmº sr. Barão de Maracajú, presidente da provincia do Amazonas, no dia 29 de março de 1879 abriu a 2ª sessão da 14ª legislatura da Assembléa Legislativa Provincial* (Manaus: Typ. do Amazonas, 1879), 40; Hemming, *Amazon Frontier*, 267–68, 272.

43. Aprigio Martins de Menezes, *Relatorio apresentado ao exm. sr. Dr. Presidente da Provincia Domingos Monteiro Peixoto pelo Dr. Aprigio Martins de Menezes, encarregado da Enfermaria dos Variolosos* (Manaus: Typographia do Commercio do Amasonas, 1874), 1–2; Hemming, *Amazon Frontier*, 268.

44. Franz Keller, *Amazon and Madeira Rivers*, 144–45, 174–75; Orbigny, *Descripción*, 362–63; Maccheti, *Diario del viaje fluvial*, 17, 19; Agassiz and Agassiz, 178–79, 306; Mathews, *Up the Amazon and Madeira Rivers*, 71–72.

45. Craig, *Recollections of an Ill-Fated Expedition*, 289–90; Mathews, *Up the Amazon and Madeira Rivers*, 54.

46. Franz Keller, *Amazon and Madeira Rivers*, 137–40; Wilkens de Mattos, *Relatorio . . . de 1869*, 9; Mathews, *Up the Amazon and Madeira Rivers*, 20; Craig, *Recollections of an Ill-Fated Expedition*, 265; Brown and Lidstone, *Fifteen Thousand Miles*, 340, 346; Fonseca, *Voyage autour du Brésil*, 287.

47. Franz Keller, *Amazon and Madeira Rivers*, 58, 143–49; Maccheti, *Diario del viaje fluvial*, 30–31; Craig, *Recollections of an Ill-Fated Expedition*, 257.

48. Franz Keller, *Amazon and Madeira Rivers*, 58, 144–45; Pinkas, *Relatorio apresentado*, 111.

49. Maccheti, *Diario del viaje fluvial*, 24, 31–33; Mathews, *Up the Amazon and Madeira Rivers*, 58–59.

50. Franz Keller, *Amazon and Madeira Rivers*, 60–62; Maccheti, *Diario del viaje fluvial*, 43; Craig, *Recollections of an Ill-Fated Expedition*, 290, 360–64, 369, 373, 392, 399, 408–09, 435.

51. Tomás M. Villavisencio to Ministro de Estado, 18 octubre 1887 (ANB MI T.237 N.63); *Boletín del Ministerio de Relaciones Esteriores y Colonización*, 15 febrero 1887, 19; Mathews, *Up the Amazon and Madeira Rivers*, 169, 171.

52. Franz Keller, *Amazon and Madeira Rivers*, 81–82, 186; Cabrera, *Navegación fluvial*, 3–6, 29; Salinas, *Navegación*, 16–17.

53. Mathews, *Up the Amazon and Madeira Rivers*, 161; Cabrera, *Navegación fluvial*, 31; Prefectura y comandancia general del Beni 1888, 6 septiembre 1888 (ANB MI T.239 N.75); Prefectura del Beni 1889, 8 mayo 1889 (ANB MI T.244 N.61).

54. Mathews, *Up the Amazon and Madeira Rivers*, 48, 135, 165.

55. Ibid., 39; Fonseca, *Voyage autour du Brésil*, 277; Salinas, *Navegación*, 17; Maccheti, *Diario del viaje fluvial*, 65.

56. Craig, *Recollections of an Ill-Fated Expedition*, 252. For Oyola's links to Baures see Limpias, *Los gobernadores de Mojos*, 280; Coimbra, *Siringa*, 51; and ANB TNC Ct Beni 2d 1902 (Fincas rústicas); and Carvalho, *Beni*, 33. As Pastor Oyola's daughter was named Cristina Oyola Ojopi, his wife was an Ojopi; see Ricardo Centeno and Patricia Fernández O., *Imágenes del auge de la goma* (La Paz: La Papelera, 1998), 18–19. Further research has failed to reveal Oyola's wife's full name, necessary to identify her as one of Hipólito Ojopi's daughters.

57. José and Francisco Keller, *Exploración del Río Madera*, 18; Salinas, *Navegación*, 17; Coimbra, *Siringa*, 7–8.

58. Franz Keller, *Amazon and Madeira Rivers*, 121; Mathews, *Up the Amazon and Madeira Rivers*, 133–34; Pinkas, *Relatorio apresentado*, 109–10.

59. Weinstein, *Amazon Rubber Boom*, 22–24; Bradford Barham and Oliver Coomes, *Prosperity's Promise* (Boulder, CO: Westview Press, 1996), 18–19; Mathews, *Up the Amazon and Madeira Rivers*, 135–36.

60. Weinstein, *Amazon Rubber Boom*, 23–24; Craig, *Recollections of an Ill-Fated Expedition*, 251.

61. Examples include Exaltación on the Madeira River, Exaltación on the Beni River, Cayuvava on the Madidi River, Concepción de Morrinhos on the Madeira

River, Carmen on the Madeira River, Trinidadcito on the Río Madre de Dios, and San Pedro on the Río Orthon.

62. Mathews, *Up the Amazon and Madeira Rivers,* 115, 160; Franz Keller, *Amazon and Madeira Rivers,* 184–85; Limpias, *Los gobernadores de Mojos,* 200.

63. Klein, 146–47; Limpias, *Los gobernadores de Mojos,* 219, 231.

64. Oficina Nacional de Inmigración, Estadística y Propaganda Geográfica, *Sinopsis estadística y geográfica de la República Boliviana, Tomo II* (La Paz: Tall. Tip.-Lit. de J.M. Gamarra, 1903), 166–67; Fifer, "Empire Builders," 127; María del Pilar Gamarra, "La participación estatal en la industria de la goma elástica," *DATA, Revista del Instituto de Estudios Andinos y Amazónicos* 4 (1993), 39–40. Bolivianos and centavos were new units of currency created in 1864 and 1870, respectively.

65. Armentia, *Diario,* 5, 25; Pinkas, *Relatorio apresentado,* 110; Cabrera, *Navegación fluvial,* 22. Hemming, *Amazon Frontier,* 277, states that the old woman mentioned by Pinkas "was to be punished by being shot from a cannon," a misinterpretation of Pinkas's "*seu patrão . . . armasse sua gente para tiral-a da canôa dos soldados*" (her patrón . . . might arm his people to pull her out of the canoe of the soldiers).

66. Rusby, *Jungle Memories,* 264–65.

67. Franz Keller, *Amazon and Madeira Rivers,* 187–95.

68. Carvalho, *Visión del Beni,* 28; Franz Keller, *Amazon and Madeira Rivers,* 186–87, 80–82; Mathews, *Up the Amazon and Madeira Rivers,* 123, 127, 129–30, 151–55, 165; Ygnacio Araúz, *Nueva vía fluvial de Bolivia* (Manaus: Typographia do "Amazonas" de A. da C. Mendes, 1868), 12.

69. Benjamín Lens, *El Departamento del Beni desde fines del 70 hasta Febrero del 71* (La Paz: Imprenta de la Union Americana—de César Sevilla, 1872) 17; Limpias, *Los gobernadores de Mojos,* 191–96; Revisita of Beni, 1861, f.219v. The three Nosas included Frutos Nosa (corregidor or cacique at various times in 1843–65), Mariano Nosa (cacique in 1862), and Juan de Mata Nosa (músico in 1861). See note 10 to chapter 2 and Revisita of Beni, 1861, f.240–240v., 248.

70. Gabriel René Moreno, *Catálogo del Archivo de Mojos y Chiquitos* (La Paz: Libreria Editorial "Juventud," 1974), 393.

71. Arteche, *JHS misión,* 15, 18–19; Gibbon, *Exploration,* 247–48.

72. Mathews, *Up the Amazon and Madeira Rivers,* 138, 161; Eric Hobsbawm, *Industry and Empire* (Harmondsworth: Penguin Books, 1968), 154.

73. Arteche, *JHS misión,* 18; Tomás M. Villavisencio to Ministro de Estado, 18 octubre 1887 (ANB MI T.237 N.63 1887).

74. Arteche, *JHS misión,* 17–18.

75. Emma Banzer Toro de Añez, *Monografía de Exaltación de la Santa Cruz* (Santa Cruz: Impresiones Offset Color "Oriente," 2004) 262; *Padron jeneral de indijenas contribuyentes de la Provincia de Mojos, Año de 1854* (ANB RV Beni 13, T. 1) f.103; Revisita of Beni, 1861 (ANB TNC RV N.10 1861), f. 46; *Matrícula jeneral del Beni para el Tribunal de Cuentas* (ANB RV Beni 1874 3) f.257; Mathews, *Up the Amazon and Madeira Rivers,* 121–22.

76. Cabrera, *Navegación fluvial,* 4–5, 29.

77. Carvalho, *Beni,* 33; Coimbra, *Siringa,* 51–52.

78. José and Francisco Keller, *Exploración del Río Madera,* 20–21; Mathews, *Up the Amazon and Madeira Rivers,* 131–32, 160.

79. Arteche, *JHS misión,* 17.

80. Actos de gobierno, correspondencia, 1841. Copiador de notas dirijidas al gobernador de la Provincia de Mojos—Año de 1841 (in Copiador de comunicaciones del año 1841), No. 44, Mayo 1 (BAHUAGRM Prefectura 1/29).

81. José Manuel Montero to Ministro del Interior, 18 junio 1846 (ANB MI T.115 N.37).

82. Franz Keller, *Amazon and Madeira Rivers*, 128; Gibbon, *Exploration*, 259; Mathews, *Up the Amazon and Madeira Rivers*, 159; Limpias, *Los gobernadores de Mojos*, 66, 139.

83. Revisita of Beni, 1861 (ANB TNC RV N.10 1861), f. 46–69; Keller, *Amazon and Madeira Rivers*, 196.

84. Keller, *Amazon and Madeira Rivers*, 186, 196.

85. Banzer Toro de Añez, *Monografía*, 261–62.

86. Araúz, 16; Gibbon, *Exploration*, 267; Maccheti, *Diario del viaje fluvial*, 3–4; Franz Keller, *Amazon and Madeira Rivers*, 79; Mathews, *Up the Amazon and Madeira Rivers*, 121; Cardús, *Las misiones franciscanas*, 290–91.

87. Cardús, *Las misiones franciscanas*, 280–81, 287–88; Cabrera, *Navegación fluvial*, 7.

88. Cardús, *Las misiones franciscanas*, 290; Suárez, *Manifiesto*, 10; Pinto, *Rumbo al Beni*, 40.

## Chapter 4

1. Suárez, *Manifiesto*, 7.
2. Ibid., 10–11.
3. Pinkas, *Relatorio apresentado*, Annexo N.3, 19.
4. Arteche, *JHS misión*, 27.
5. Antelo, "Sucesos del Beni," 3–4; Suárez, *Manifiesto*, 4; Arteche, *JHS misión*, 22, 29.
6. Hernando Sanabria Fernández, *En busca de Eldorado* (Santa Cruz: Universidad Gabriel René Moreno, Buenos Aires: Imprenta López, 1958), 209–10.
7. Marcelino Marañón to Ministro de Gobierno, 9 julio 1888 (ANB MI T.239 N.75); José María Urdininea to Ministro de Gobierno, 23 enero 1888 (ANB MI T.239 N.75).
8. Cardús, *Las misiones franciscanas*, 288–89, n.1.
9. Antelo, "Sucesos del Beni," 2.
10. Ibid., 1–2.
11. Ibid.
12. Ibid., 2.
13. Ibid.
14. Ibid.
15. Ibid.
16. Ibid.
17. Ibid.
18. Ibid., 2, 4.
19. Ibid., 2; Revisita of Beni, 1861 (ANB TNC Revisitas 10), f. 228; Revisita of Beni, 1867 (ANB TNC RV 8); *Matrícula jeneral del Beni para el Tribunal de Cuentas* (1874); "Gacetilla," *Los Debates*, 5 mayo 1887, 3.

20. Becerra, *Reliquias de Moxos*, 260; Rafael Arteaga Terrazas, quoted in Ignacio Callaú Barbery, "La Guayochería," *Revista de la Universidad Autonoma "Gabriel René Moreno"* 3, no. 8 (Dic. 1950): 179; Revisita of Beni, 1867 (ANB TNC RV 8).

21. Antelo, "Sucesos del Beni," 2.

22. Ibid.

23. "Gacetilla," 3.

24. Antelo, "Sucesos del Beni," 2–3.

25. Ibid., 3.

26. Ibid.

27. Ibid.

28. Ibid.

29. Ibid.

30. Ibid.

31. Ibid.

32. Suárez, *Manifiesto*, 11.

33. Orbigny, *Descripción*, 194, 291, 322–23.

34. Coimbra, *Siringa*, 43.

35. Rafael Arteaga Terrazas, quoted in Callaú, "La Guayochería," 179.

36. Coimbra, *Siringa*, 43.

37. "Gacetilla," 3; Riester, *En busca*, 318, 320, 323–24.

38. Suárez, *Manifiesto*, 11.

39. Stephen Hugh-Jones, "Shamans, Prophets, Priests, and Pastors," in Nicholas Thomas and Caroline Humphrey, eds. *Shamanism, History, and the State* (Ann Arbor: University of Michigan Press, 1994), 32, 47–50.

40. *Sublevación de los Indios Chiriguanos en las provincias de Azero y Cordillera, pertenecientes a los Departamentos de Sucre y Santa-Cruz de la República de Bolivia* (Potosí: Imp. de "El Porvenir," 1892).

41. Vincent Crapanzano, "Spirit Possession," in Mircea Eliade, ed., *The Encyclopedia of Religion* (New York: MacMillan, 1987), 12–13.

42. For a Mojo account of soul flight, see Riester, *En busca*, 322–23.

43. Anna-Leena Siikala, "Shamanism: Siberian and Inner Asian Shamanism," in Mircea Eliade, ed., *The Encyclopedia of Religion* (New York: MacMillan, 1987), 213–14.

44. Other than the case of Guayocho, examples and references include: Quechua and Aymara in Julian H. Stewart, ed., *Handbook of South American Indians*, Vol. 2 (Washington, DC: United States Government Printing Office, 1946), 469, 565–66; Aymara in Hans Buecher and Judith-Maria Buecher, *The Bolivian Aymara* (New York: Holt, Rinehart and Winston, 1971), 96; Yucatan Maya and Chiapas Maya in Nelson Reed, *The Caste War of Yucatan* (Stanford: Stanford University Press, 1964), 134–40, 161, 182–83; Taino in Birgit Faber Morse, "The Salt River Site, St. Croix, at the Time of the Encounter," in Samuel Wilson, ed., *The Indigenous People of the Caribbean* (Gainesville: University Press of Florida, 1997), 40; Cree, Saulteaux, and Cheyenne in Eliade, *Shamanism*, 255, 335; Chukchi in Siikala, "Shamanism," 214; Tepehuan in Edward Spicer, *Cycles of Conquest* (Tucson: University of Arizona Press, 1962), 28; Rukuba in Jean-Claude Muller, "La pierre qui parle ou le savoir ventriloque," *Anthropologie et sociétés* 17, no. 3 (1993): 93–101; Yoruba and Age in Robert Thompson,

*Flash of the Spirit* (New York: Vintage Books, 1984), 43–44; Zulu, Maori, Eskimo, and Patagonian in "Ventriloquism," in *New Encyclopaedia Britannica*, 15th ed., Vol. 12 (Chicago: Encyclopaedia Brittanica, 1997), 309, and "Ventriloquism," in *Collier's Encyclopedia*, Vol. 23 (New York: P. F. Collier, 1995), 79. These last two references also mention possible ventriloquism among ancient Egyptians, Greeks, and Hebrews, medieval Europeans, Chinese, and Asian Indians. In some of the African examples, ventriloquism developed as a response to or as a complement to the dominant method of spirit possession.

45. Reed, *Caste War*, 134–40, 161, 182–83.

46. Apter offers this interpretation of African spirit possession as discursive agency in Apter, *Beyond Words*, 9.

47. See Riester, *En busca*, 311–15; Jorge Cortés Rodríguez's notes in Arteche, *JHS misión*, 39–42; and Lehm, *Milenarismo*, 44–48, 57.

48. See Cortés's notes in Arteche, *JHS misión*, 42, and Lehm, *Milenarismo*, 58.

49. Norman Cohn, *The Pursuit of the Millennium* (New York: Oxford University Press, 1970); Brown, "Beyond Resistance," 292–93; Jones, "Conflict between Whites and Indians," 291–93.

50. Suárez, *Manifiesto*, 11.

51. Arteche, *JHS misión*, 27.

52. Rafael Artega Terrazas, "La guayochería la inventaron las autoridades," *La Flecha*, 18 noviembre 1942, quoted in Cortés's notes in Arteche, *JHS misión*, 34–35. Cortés himself partly agrees with this opinion in the same pages.

53. Antelo, "Sucesos del Beni," 3.

54. Ibid.; Suárez, *Manifiesto*, 3; Daniel Suárez to Ministro de Gobierno, 19 febrero 1887 (ANB MI T.234 N.54).

55. Antelo, "Sucesos del Beni," 3; Suárez, *Manifiesto*, 2.

56. Antelo, "Sucesos del Beni," 3.

57. Arteche, *JHS misión*, 28.

58. Antelo, "Sucesos del Beni," 3.

59. Ibid.

60. Ibid.

61. Arteche, *JHS misión*, 28.

62. Antelo, "Sucesos del Beni," 3.

63. Ibid., 3–4.

64. Antelo, "Sucesos del Beni," 4; Arteche, *JHS misión*, 28; "Alarmantes noticias," *La Estrella del Oriente*, 23 abril 1887, 2.

65. Antelo, "Sucesos del Beni," 4.

66. Arteche, *JHS misión*, 29.

67. Antelo, "Sucesos del Beni," 4.

68. Ibid.

69. "Alarmantes noticias," 2; Daniel Suárez to Prefect of Chuquisaca, 26 marzo 1887 (CBDHCh SCP 1887 Caja 1).

70. Suárez, *Manifiesto*, 3; Daniel Suárez to Ministro de Gobierno, 28 marzo 1887 (ANB MI T.234 N.54).

71. Antelo, "Sucesos del Beni," 4.

72. Ibid.; Arteche, *JHS misión*, 28.

73. Antelo, "Sucesos del Beni," 4; Arteche, *JHS misión*, 28–29.

74. Antelo, "Sucesos del Beni," 4.

75. Arteche, *JHS misión*, 29; Suárez, *Manifiesto*, 4.

76. Arteche, *JHS misión*, 29–30; Antelo, "Sucesos del Beni," 4; Suárez, *Manifiesto*, 4.

77. Suárez, *Manifiesto*, 4; Becerra, *Reliquias de Moxos*, 261; Callaú, "La Guayochería," 180–81.

78. Daniel Suárez to Ministro de Gobierno, 1 julio 1887 (ANB MI T.234 N.54).

79. Arteche, *JHS misión*, 28; "Acta," *La Estrella del Oriente*, 1 septiembre 1887, 2.

80. "Acta," 2.

81. Daniel Suárez to Ministro de Gobierno, 4 julio 1887 (ANB MI T.234 N.54).

82. "Oficial: Gobierno," *Los Debates*, 15 mayo 1887, 1; "Oficial: Gobierno," *Los Debates*, 2 junio 1887, 2–3.

83. Carpio to Prefect of Santa Cruz, 7 mayo 1887 (BAHUAGRM Prefectura 3–116); "Gacetilla," *Los Debates*, 29 abril 1887, 5; "Boletín del día" and "Oficial: Gobierno," *Los Debates*, 15 mayo 1887, 1; "Gacetilla," *Los Debates*, 11 junio 1887, 4.

84. Suárez, *Manifiesto*, 7.

85. Carpio to Prefect of Santa Cruz, 28 mayo 1887 (BAHUAGRM Prefectura 3–116); "Oficial: Gobierno," *Los Debates*, 17 junio 1887, 3; Urdininea to Prefect of Chuquisaca, 18 enero 1887, 22 abril 1887, 23 abril 1887, 26 abril 1887 (CBDHCh SCP 1887 Caja 1); Urdininea to Prefect of Chuquisaca, 4 abril 1887, 27 mayo 1887 (CBDHCh SCP 1887 Caja 2); "Oficial: Gobierno" and "Gacetilla," *Los Debates*, 7 junio 1887, 3–4; "Boletín del día," *Los Debates*, 8 junio 1887, 1.

86. Arteche, *JHS misión*, 9–12.

87. Ibid., 14–15; "Acta," 2.

88. Arteche, *JHS misión*, 17.

89. Ibid., 19; Urdininea to Ministro de Gobierno, 10 septiembre 1887 and 17 septiembre 1887, published under "Gobierno," *Los Debates*, 29 octubre 1887, 2.

90. Urdininea to Ministro de Gobierno, 10 septiembre 1887 and 17 septiembre 1887, published under "Gobierno," *Los Debates*, 29 octubre 1887, 2; Suárez, *Manifiesto*, 1–11.

91. Arteche, *JHS misión*, 20.

92. Ibid., 20–21.

93. Ibid., 21. For full accounts of the Loma Santa movement, see Riester, *En busca*, and Lehm, *Milenarismo*. Both authors regard Guayocho's movement as an earlier manifestation of Loma Santa ideology, as does Ana María Lema, comp., *Pueblos indígenas de la Amazonia boliviana* (La Paz: CID-Plural, 1998), 11.

94. Arteche, *JHS misión*, 21–22.

95. Ibid., 22.

96. Ibid.

97. Ibid. 22–23.

98. Ibid., 23.

99. Ibid., 23–24; Marcelino Marañón to Ministro de Gobierno, 9 julio 1888 (ANB MI T. 239 N. 75).

100. Arteche, *JHS misión*, 24.

101. Ibid., 24–25.

102. Ibid., 25–26.

103. Ibid., 25–26; Revisita of Beni, 1861 (ANB TNC RV 10), f. 227; Revisita of Beni, 1867 (ANB TNC RV 8), f. 42.

104. Arteche, *JHS misión*, 25–26.

105. Ibid., 26.

## Chapter 5

1. "Sensible pérdida," *La Estrella del Oriente*, 26 noviembre 1887, 2.

2. Simón Velasco to Prefect of Beni, 14 febrero 1892 with note by Prefect Samuel González Portal, 23 febrero 1892, signed by Notario de Hacienda Miguel Antelo Quiton (ANB MI T. 264 N. 60); *Registro de las fincas rusticas ó agricolas*, Provincia del Ytenes, Cantón Báures (ANB TNC Beni 2d 1902, Fincas rusticas); *La Democracia*, 12 agosto 1906, 4.

3. "Dr. José María Urdininea," *La Democracia*, 18 septiembre 1909, 3.

4. José María Urdininea to Ministro de Gobierno, 23 enero 1888 (ANB MI T.239 N.75).

5. Marcelino Marañón to Ministro de Gobierno, 9 julio 1888 (ANB MI T.239 N.75).

6. Santos M. Justiniano to Ministro de Gobierno, 31 diciembre 1888 (ANB MI T.237 N.75).

7. *Archivo del Tribunal Nacional de Cuentas. Seccion tercera. Comprende de 1890 a 1906. Departamento Beni. Número 51 bis. Año 1902. Catastro de la Provincia de Mojos* (ANB TNC Ct Beni 2d 1902 Fincas rusticas).

8. "Peticiones de terreno," *El Eco del Beni*, 26 diciembre 1912, 6.

9. *Beni-Bolivia. Catastro de la provincia del Cercado i su capital. 1925* (ANB TNC Ct Beni 1x 1925), f.27, 63, 65.

10. Santos M. Justiniano to Ministro de Gobierno, 31 diciembre 1888 (ANB MI T.237 N.75).

11. Enrique Jordá et al., eds., *Historia del pueblo Mojo* (La Paz: Talleres Gráficos de Editorial Popular, 1989), 78.

12. Ibid., 78–79.

13. *Beni-Bolivia. Catastro de la provincia del Cercado i su capital. 1925* (ANB TNC Ct Beni 1x 1925), f.27–28, 65–67.

14. Jordá, *Historia del pueblo Mojo*, 78–79.

15. "Capellanía o doctrina de San Lorenzo de Mojos," *Archivo de la Comisaria Franciscana de Bolivia* (Mayo 1919): 301.

16. *Matrícula Jeneral del Beni para el Tribunal de Cuentas* (ANB RV Beni 1874 3), 108.

17. Jordá, *Historia del pueblo Mojo*, 77.

18. Central de Pueblos Indígenas del Beni, Centro de Información y Documentación para el Desarrollo del Beni, *Hacia una propuesta indígena de descentralización del estado* (La Paz: PROADE: ILDIS, 1995), 156–57.

19. Pinto, *Rumbo al Beni*, 43.

20. Zacarias Ducci, *Diario de la visita a todas las misiones existes en la República de Bolivia—América Meridional* (Asís: Tipografía de la Porciúncula, 1895), 90.

21. "El volumoso Buey del Apocalipsis, el gran Apis," *El Correo del Beni*, 18 septiembre 1894, 2, which mentions "the land of Guayocho and Santos Noco."

22. Luis Carlos Paravicini, *Diccionario legislativo boliviano, Tomo II* (La Paz: Editorial Agiurre, 1991), 1154.

23. Fr. Bernardino Pesciotti to Ministro de Relaciones Exteriores y Culto, 7 enero 1901, in *Informe que presenta al Supremo Gobierno el R.P. Prefecto Fr. Bernardino J. Pesciotti sobre el movimiento de las misiones de su cargo* (Santa Cruz: Tipografía Industrial, 1914), Anexo N° 1°, iii–iv.

24. *Censo general de la población de la República de Bolivia*, Vol. 1 (1902 edition), 25, 328; Vol. 2 (1973 reprint), Censo de la población en 1900, 19.

25. Pesciotti to Ministro de Relaciones Exteriores y Culto, 7 enero 1901, iii.

26. Fray Bernardino Pesciotti, "Informe que eleva al Supremo Gobierno el P. Prefecto de las Misiones de Guarayos," *Crónica Guaraya*, N° 58 (11 febrero 1918): 540n.

27. José Santos Noco, "INSERCIONES. San Antonio–Trinidad, julio 3 de 1905. PARA "LA DEMOCRACIA." Trinidad," *La Democracia*, 9 julio 1905, 3; M. Pacífico Roca, "LITERATURA. 'El destino de la raza india,'" *La Democracia*, 2 julio 1905, 2.

28. Santos Noco, "INSERCIONES," 3.

29. "Importante construcción," *La Democracia*, 3 junio 1906, 4; "Casas de gobierno," *La Democracia*, 12 mayo 1907, 4.

30. "Inundación," *La Democracia*, 9 febrero 1907, 4.

31. Erland Nordenskiöld, *Indianer und Weisse in Nordostbolivien* (Stuttgart: Strecker und Schröder, 1923), 72–73.

32. Bernardino Pesciotti, "Itinerario al Beni," *Archivo de la Comisaria Franciscana de Bolivia*, Año V, N° 55 (Julio 1913): 220–21.

33. Pesciotti, "Itinerario al Beni," 221; Fray Santiago Mendizabal, *Vicariato Apostólico del Beni* (La Paz: Imp. "Renacimiento," 1932), 255; "Informe del R.P. Francisco Pierini (Conclusión)," *Archivo de la Comisaria Franciscana de Bolivia*, Año IV, N° 41 (Mayo 1912): 149–50.

34. "Informe del R.P. Francisco Pierini (Conclusión)," 150.

35. "Social," *El Eco del Beni*, 20 junio 1912, 5.

36. Fray Francisco Pierini to Fiscal Angel Lara, 2 julio 1912, transcribed in Fiscal Angel Lara to Prefect of Beni, 15 julio 1912, and reprinted in "Documentos de la fiscalia del distrito," *El Eco del Beni*, 1 agosto 1912, 6; "La misión de San Lorenzo," *El Eco del Beni*, 1 agosto 1912, 4.

37. "Ley de 26 de octubre," in *Anuario de leyes, decretos, resoluciones y ordenes supremas. Año de 1905 (edición oficial)* (La Paz: Tipografía Artística de Castillo y C$^{a}$, 1906), 433–34.

38. Jones, "Conflict between Whites and Indians," 80.

39. The correlation between the collapse in rubber prices and an increase in land petitions in the Province of Iténez is made explicit in "De provincias," *El Eco del Beni*, 14 agosto 1913, 1.

40. By March 1913, carayanas had filed at least two land petitions south of San Ignacio, in the direction of San Lorenzo. The earlier of the two was for a place named Maguncia, which is not far to the north of San Lorenzo and San Francisco. See "Solicitudes de terreno," *El Eco del Beni*, 14 septiembre 1913, 9–10.

41. "La misión de San Lorenzo," *El Eco del Beni*, 1 agosto 1912, 4.

42. "Por el cantón San Ignacio," *El Eco del Beni*, 15 agosto 1912, 3.

43. "Breves apuntes sobre la misión de San Lorenzo," *El Eco del Beni*, 29 agosto 1912, 3–4.
44. Ibid.
45. Ibid.
46. Ibid.
47. "Resolución suprema de 30 de agosto," in República de Bolivia, *Anuario de leyes, decretos y resoluciones supremas de 1912*, compilado por Mariano Rivas L., Tomo 1 (La Paz: Talleres Gráficos "La Prensa," 1913), 677–78.
48. "La nueva misión en el Mamoré," *El Eco del Beni*, 17 octubre 1912, 3.
49. Ibid.
50. "La nueva misión en el Mamoré, II," *El Eco del Beni*, 24 octubre 1912, 3.
51. Méndez Roca, Antezana, and Otazo to Ministro de Guerra y Colonización, 10 diciembre 1912, and Ministro de Guerra y Colonización to Méndez Roca, Antezana, and Otazo, 14 diciembre 1912, reprinted in "Una misión de frailes," *El Eco del Beni*, 30 enero 1913, 6.
52. Fray Francisco Pierini to Ministro de Colonización y Guerra, 29 diciembre 1912, reprinted in *Archivo de la Comisaria Franciscana de Bolivia*, Año V, Nº 49 (Enero 1913): 5–6.
53. Ibid., 7–12.
54. Fr. Bernardino Pesciotti to Ministro de Estado en el Despacho de Guerra y Colonización, 31 diciembre 1912, in *Archivo de la Comisaria Franciscana de Bolivia*, Año V, Nº 52 (Abril 1913): 106–08.
55. "La misión de San Lorenzo," *El Eco del Beni*, 16 enero 1913, 4.
56. "De San Lorenzo," *La Democracia*, 1 febrero 1913, 4.
57. Fr. Rufino Holler and Fr. Juan Félix Jenewein to Fr. Francisco Pierini, 6 enero 1913, in *Archivo de la Comisaria Franciscana de Bolivia*, Año V, Nº 52 (Abril 1913): 111–12; "De San Lorenzo," 4.
58. Holler and Jenewein to Pierini, 6 enero 1913, 112–13.
59. Ibid., 112.
60. Ibid., 113.
61. Ibid., 113–14.
62. Ibid., 111–12.
63. Fr. Bernardino Pesciotti to Ministro de Estado en el Despacho de Guerra y Colonización, 31 diciembre 1913, in Pesciotti, *Informe que presenta al Supremo Gobierno el R.P. Prefecto Fr. Bernardino J. Pesciotti sobre el movimiento de las misiones de su cargo*, 3–4.
64. "El camino a Trinidad," *Archivo de la Comisaria Franciscana de Bolivia*, Año VII, Nº 81 (Septiembre 1915): 295–96.
65. Fr. Bernardino Pesciotti to Ministro de Estado en el Despacho de Guerra y Colonización, 31 diciembre 1916, reprinted as "Informe que eleva al Supremo Gobierno el P. Prefecto de las Misiones de Guarayos," *Crónica Guaraya*, Nº 58 (11 febrero 1918): 541.
66. Ibid.
67. Ibid., 540–41.
68. Ibid., 541–42; Paravicini, *Diccionario*, 1161.
69. "Capellanía o doctrina de San Lorenzo en Mojos," 301.
70. Ibid., 301–2.

71. "Capellanía o doctrina de San Lorenzo de Mojos," 301; "Rdo. P. Pedro Celestino Herritsch," *Archivo de la Comisaria Franciscana de Bolivia,* Año XI, N° 126 (Junio 1919): 357–58.

72. "La Vicaría Apostólica del Beni," *Archivo de la Comisaria Franciscana de Bolivia,* Año X, N° 118 (Octubre 1918): 305–15; "Sección de actualidad," *Archivo de la Comisaria Franciscana de Bolivia,* Año XII, N° 133 (Enero 1920): 35–36.

73. Pinto, *Rumbo al Beni,* 39–45.

74. *Beni-Bolivia. Catastro de la provincia del Cercado i su capital. 1925* (ANB TNC Ct Beni 1x 1925); *Libro de actas y declaraciones* (ANB TNC Ct Beni 1 1924), 109–10.

75. *Beni-Bolivia. Catastro de la provincia del Cercado i su capital. 1925* (ANB TNC Ct Beni 1x 1925).

76. Federico Román, *Ferrocarril Cochabamba-Santa Cruz. Razones en pro del trazo del Norte* (La Paz: Escuela Tipográfica Salesiana, 1922), 9–10.

77. *Libro de actas y declaraciones* (ANB TNC Ct Beni 1 1924), 111.

78. Carvalho, *Visión del Beni,* 36; Pinto, *Rumbo al Beni,* 43; Arnaldo Lijeron Casanovas, *Facetas de la cultura beniana* (Trinidad: Talleres de la Oficina Técnica de Poligrafiados, 1982), 96.

## Chapter 6

1. See ANB MI T.244 N.61, Prefectura del Beni 1889, lists of the Columna de Celadores for 22 enero 1889, 4 febrero 1889, 23 marzo 1889, and 8 abril 1889; of the Columna del Orden for 23 mayo 1889 and 15 junio 1889; and of the Columna Mamoré N° 10 for 10 agosto 1889.

2. Manuel Limpias P. to Fiscal de Santa Cruz, 9 enero 1893, in "Oficial," *El Correo del Beni,* 18 septiembre 1894, 1.

3. Fabián Vaca Chávez, "Por el Oriente y por el noroeste de la república. Problemas e iniciativas," *El Eco del Beni,* 28 marzo 1912, 3; Raul Monje Roca, *El río Mamoré* (La Paz: Instituto Boliviano de Cultura, 1977), 29–31.

4. *Anuario de leyes y supremas disposiciones de 1896* (La Paz: Imprenta de "El Comercio," 1898), 132–34.

5. *Anuario de leyes y supremas disposiciones de 1897* (n.d.), 63–68.

6. "Sobre las noticias de 'Gironda,'" *El Eco del Beni,* 20 febrero 1913, 4.

7. "Guerra a muerte," *La Democracia,* 11 febrero 1906, 2.

8. "Aviso," *La Democracia,* 11 febrero 1906, 1.

9. M. Pradovelo, "Correspondencias," *La Democracia,* 29 abril 1906, 4.

10. Isaiah Bowman, "Geographical Aspects of the New Madeira-Mamoré Railroad," *Bulletin of the American Geographical Society* 45, No. 4 (1913): 278; *Anexos a la memoria presentada en la legislatura de 1907* (La Paz: Tip. Artística de Castillo y Cª, 1908), 105–06; Schurz, "Distribution of Population," 216.

11. Jones, "Conflict between Whites and Indians," 121, 124–29.

12. "Correspondencias," *La Democracia,* 17 octubre 1908, 2; "Decreto Supremo," *El Eco del Beni,* 10 octubre 1912, 5; "Via de Riberalta a Exaltacion," *La Voz del Pueblo,* 18 diciembre 1908, 1.

13. "Abolición necesaria," *La Democracia,* 21 julio 1907, 1.

14. "Urjente rectificación del catastro," *La Democracia,* 12 junio 1913, 2.

15. "La industria pecuaria," *La Democracia*, 20 octubre 1907, 2.

16. "De Provincias," *El Eco del Beni*, 14 agosto 1913, 1.

17. "Peticiones," *La Democracia*, 6 octubre 1907, 3; "Peticiones," *La Democracia*, 26 enero 1908, 4; "Peticiones," *La Democracia*, 15 agosto 1908, 1; "Testimonio," *El Eco del Beni*, 20 febrero 1913, 1.

18. Lehm, "Proyecto," 208–11.

19. "Lo que dice "El Tiempo" de La Paz. Las tierras baldías. Abusos que se cometen en el Beni," *La Democracia*, 10 febrero 1912, 2.

20. "Oficial," *El Eco del Beni*, 1 agosto 1912, 6.

21. "Peticiones de terreno," *El Eco del Beni*, 1 agosto 1912, 8.

22. Lehm, "Proyecto," 208–11.

23. "Peticiones de terreno," *El Eco del Beni*, 25 julio 1912, 7; "Peticiones de terreno," *El Eco del Beni*, 7 marzo 1912, 1; "Solicitudes de terrenos," *El Eco del Beni*, 23 octubre 1913, 11.

24. "Peticiones de terreno," *El Eco del Beni*, 30 mayo 1912, 7; "Peticiones de terreno," *El Eco del Beni*, 25 julio 1912, 7; "Peticiones," *La Democracia*, 12 septiembre 1908, 1.

25. "Lluvia," *La Democracia*, 20 enero 1907, 4; "El agua avanza," *La Democracia*, 10 febrero 1912, 4; "Por las industrias del país." *La Democracia*, 9 marzo 1912, 1.

26. "Prevención," *La Voz del Pueblo*, 27 noviembre 1926, 4; "Santa Rosa," *La Voz del Pueblo*, 20 julio 1921, 1.

27. Lehm, Lijerón, and Vare, *Diagnóstico socio-económico de los indígenas mojeños*, 13–14; Langer, *Economic Change*, 21; *Censo general de la población de la República de Bolivia*, Vol. 2 (1973 reprint), lxix–lxxi.

28. "Nómina de mozos matriculados en el pte. mes," *El Eco del Beni*, 4 septiembre 1913, 7.

29. "El estado y el Beni," *La Democracia*, 21 agosto 1904, 2.

30. "En San Ignacio," *El Eco del Beni*, 20 febrero 1913, 3–4.

31. "Urjente rectificación del catastro," *La Democracia*, 12 julio 1913, 2.

32. "Jurisprudencia," *La Democracia*, 29 octubre 1905, 2; "Remitidos," *La Democracia*, 14 diciembre 1905, 2; "Un buen corregidor," *La Democracia*, 3 junio 1906, 4; and "Contestación," *La Democracia*, 10 junio 1906, 4. For carayana competition over cattle, see "Remitido," *La Democracia*, 2 noviembre 1905, 2; "Avisos," *La Democracia*, 21 octubre 1906, 4; and "Un hacendado de hurinas," "Bártulo o malicia" and "El abigeato," all in *La Democracia*, 28 octubre 1906, 4.

33. "Condiciones de vida," *La Democracia*, 22 octubre 1905, 1.

34. Heath, "Camba," 53–55.

35. República de Bolivia, Ministerio de Hacienda, *Características demográficas de la población en Bolivia* (La Paz: 1997), 47; Alvaro Díez Astete and David Murillo, *Pueblos indígenas de tierras bajas* (La Paz: Ministerio de Desarrollo Sostenible y Planificación, Viceministerio de Asuntos Indígenas y Pueblos Originarios, Programa Indigenista-PNUD, 1998), Cuadro #23.

36. Jordá, *Historia del pueblo Mojo*, 80–81.

37. Lehm, Lijerón, and Vare, *Diagnóstico socio-económico de los indígenas mojeños*, 13.

38. Jones, "Conflict between Whites and Indians," 126.

39. "Rómulo Suárez," *La Democracia*, 19 abril 1908, 3; "La delincuencia aumenta. El crimen de La Loma," *La Democracia*, 26 abril 1908, 2; "Páginas de sangre. La criminalidad aumenta," *La Democracia*, 10 octubre 1908, 2; "Horrible asesinato," *La Democracia*, 10 octubre 1908, 4; "Las Pascuas," *La Democracia*, 29 diciembre 1907, 4; "En San Ignacio," *El Eco del Beni*, 20 febrero 1913, 4; "Sobre las noticias de 'Gironda,'" *El Eco del Beni*, 20 febrero 1913, 4.

40. Samuel González Portal to Ministro de Gobierno y Colonización, 23 enero 1892 (ANB MI T. 264 N. 60); Jones, "Conflict between Whites and Indians," 119, 143 n.27.

41. "Social," *El Eco del Beni*, 1 agosto 1912, 4; "Por el cantón San Ignacio," *El Eco del Beni*, 15 agosto 1912, 3; "En San Ignacio. Un corregidor modelo," *El Eco del Beni*, 20 febrero 1913, 3; Orbigny, *Descripción*, 254–55.

42. Libro de actas y declaraciones (ANB Ct Beni 1 1924), 62–67.

43. Jones, "Conflict between Whites and Indians," 130–31.

44. Libro 8º de catastro de ganado vacuno. Provincia Mojos. 1901–1902 duplicado (ANB TNC Ct Beni 2c 1902 Ganado vacuno); Archivo del Tribunal Nacional de Cuentas. Sección tercera. Comprende de 1890 a 1906. Departamento Beni. Número 51 bis. Año 1902. Catastro de la Provincia de Mojos (ANB TNC Ct Beni 2d 1902 Fincas rústicas); Beni-Bolivia. Catastro de la provincia del Cercado i su capital. 1925 (ANB TNC Ct Beni 1x 1925).

45. Jones, "Conflict between Whites and Indians," 131–32, 156–62, 164–68; Lehm, *Milenarismo*, 24; Xavier Albó et al., *Para comprender las culturas rurales en Bolivia* (La Paz: MEC-CIPCA-UNICEF, 1989), 216.

46. "Abusos," *La Democracia*, 26 octubre 1905, 2.

47. "Remitidos," *La Democracia*, 14 diciembre 1905, 2. Gil continued to abuse his authority in 1906 by robbing cattle from two carayana patrones; see "Curioso caso," *La Democracia*, 22 abril 1906, 4.

48. "Inícuo," *La Democracia*, 19 abril 1908, 3; "Buena medida," *La Democracia*, 30 mayo 1908, 4.

49. "Transcripción, 'Trinidad,'" *La Democracia*, 29 febrero 1908, 4.

50. "Matrícula de carretones" and "Matrícula de cachorros," *La Democracia*, 1 febrero 1913, 2.

51. "Policia," *El Eco del Beni*, 17 julio 1913, 5; "Policia," *El Eco del Beni*, 14 agosto 1913, 5.

52. "La institución municipal en Trinidad," *La Democracia*, 13 enero 1907, 2.

53. "Vagos y mal entretenidos," *El Eco del Beni*, 4 septiembre 1913, 6; "Contra el alcoholismo," *El Eco del Beni*, 2 octubre 1913, 7.

54. "La policia," *La Democracia*, 3 febrero 1912, 4.

55. "El peligro de los carretones," *El Eco del Beni*, 18 septiembre 1913, 6.

56. Paravicini, *Diccionario*, Tomo II, Apendice, 20.

57. "Nómina de los contribuyentes al impuesto de Prestación Vial de esta capital que han pagado en dinero a razon de Bs. 6-c/u. correspondiente al presente año de 1913," *El Eco del Beni*, 14 agosto 1913, 2, and 4 septiembre 1913, 2–3; "Nómina de los contribuyentes al impuesto de Prestación Vial de esta Capital y Cercado, que han pagado en trabajo personal el presente año 1913," *El Eco del Beni*, 18 septiembre 1913, 3–4.

58. "De la Prefectura" and "La prestación vial en el Cercado," *El Eco del Beni*, 23 octubre 1913, 2–3 and 5 respectively.

59. Paravicini, *Diccionario*, Tomo II, 832–33; Luis Velasco Crespi, "Fuerzas Armadas," in Josep Barnadas, ed., *Diccionario histórico de Bolivia*, I (Sucre: Grupo de Estudios Históricos, 2002), 895.

60. Articles in *La Democracia* entitled "Inserción," 5 noviembre 1905, 2; 9 noviembre 1905, 1; 18 noviembre 1905, 1; and "Sorteo," 7 diciembre 1905, 2.

61. "Sorteo," *La Democracia*, 16 diciembre 1906, 4.

62. "Abusos," *La Democracia*, 28 enero 1906, 4.

63. "Conscripción militar," *La Democracia*, 16 noviembre 1905, 2; "El camino San Borja-Chulumani," *El Eco del Beni*, 20 septiembre 1930, 5.

64. "Provincias," *El Eco del Beni*, 16 octubre 1913, 8; Cámara de Comercio e Industria Trinidad, *Informe de los delegados del Departamento del Beni, que asistieron al primer congreso nacional de agricultura y ganaderia* (La Paz: Editorial "America," 1931), 10–11.

65. "Acto de justicia," *La Democracia*, 22 enero 1905, 2.

66. "Abusos," *La Democracia*, 18 febrero 1906, 4. Later in 1906, Cortéz continued to abuse Indians, though he was no longer corregidor. See "Ecos de Loreto," *La Democracia*, 16 diciembre 1906, 4.

67. "Terrenos urbanos," *El Eco del Beni*, 13 septiembre 1919, 2.

68. Lehm, Lijerón, and Vare, *Diagnóstico socio-económico de los indígenas mojeños*, 16–17; 45, n.30.

69. Lehm, "Proyecto," 211; Lehm, Lijerón, and Vare, *Diagnóstico socio-económico de los indígenas mojeños*, 16, 45 n.29.

70. "Las Pascuas," *La Democracia*, 29 diciembre 1907, 4.

71. Lehm, Lijerón, and Vare, *Diagnóstico socio-económico de los indígenas mojeños*, 17.

72. Jones, "Conflict between Whites and Indians," 118, 130, 147–48.

73. "El Beni y la instrucción," *La Democracia*, 14 octubre 1906, 2; "Pruebas escolares," *La Democracia*, 27 octubre 1907, 4; "Exámenes," *La Democracia*, 1 noviembre 1907, 4.

74. "Presupuesto del Departamento del Beni–Gestión 1912," *La Democracia*, 10 febrero 1912, 2; "Escuela municipal de indígenas," *El Eco del Beni*, 23 octubre 1913, 7.

75. "Escuela parroquial," *El Eco del Beni*, 13 septiembre 1930, 4, 8.

76. "Festividad," *La Democracia*, 18 junio 1905, 4.

77. "Festividad," *La Democracia*, 10 junio 1906, 4; "La fiesta," *La Democracia*, 13 junio 1908, 4.

78. "La fiesta local," *El Eco del Beni*, 23 mayo 1912, 5.

79. "Abuso," *La Democracia*, 17 junio 1906, 2.

80. "La fiesta," *La Democracia*, 25 junio 1905, 4.

81. "Fiestas civicas," *El Correo del Beni*, 8 septiembre 1895, 1.

82. "Fiestas patrias," *La Democracia*, 12 agosto 1906, 3.

83. "Las fiestas patrias," *El Eco del Beni*, 14 agosto 1913, 5.

84. "Ecos de Villa Bella," *La Democracia*, 27 enero 1912, 4.

85. "Las campanas de Mojos," *El Eco del Beni*, 14 agosto 1913, 5.

86. "Lutero con calzones y en mangas de camisa," *La Democracia,* 29 enero 1905, 2.

87. "Por la higiene," *El Eco del Beni,* 15 agosto 1912, 4.

88. "Municipalidad," *El Eco del Beni,* 2 octubre 1913, 9.

89. "Remitido" and "En descubierto," *La Democracia,* 1 julio 1906, 4; "Remitidos," *La Democracia,* 8 julio 1906; and "Remitidos," *La Democracia,* 5 agosto 1906, 3.

90. Jones, "Conflict between Whites and Indians," 120.

91. Lehm, Lijerón, and Vare, *Diagnóstico socio-económico de los indígenas mojeños,* 45–46, n.38.

## Conclusion

1. María Elena García, *Making Indigenous Citizens* (Stanford: Stanford University Press, 2005), 165.

# *Glossary*

**abadesa:** a Mojo female who performs services for the church

**Acanga Piranga:** a Tupí-Guaraní-speaking indigenous group from the highlands east of the upper Madeira River, also called Uru-Eu-Wau-Wau

**achiaco:** precontact name for a Mojo chief

**agregado:** a person "added on" to a settlement without being considered part of it

**aguardiente:** cane liquor

**alcalde de campo:** one title for a government official heading a vice cantón

**alguacil del juzgado de letras:** bailiff of the court

**almendra** (<Spanish **almendra,** almond): Brazil nut

**Araona:** a Tacanan-speaking indigenous group of the northern rainforest; lived independently north of the Río Madre de Dios

**Arara:** an indigenous group of the Madeira River

**arroba:** a unit of weight equal to about 11.3 kilograms (25 pounds)

**ausente:** an absent tributary

**bajones:** huge panpipes made from rolled palm leaves

**bárbaros (indios bárbaros):** Spanish term for independent, non-Christian Indians living outside of state control

**barraca** (<Brazilian Portuguese **barracão**): rubber extraction settlement

**bastón:** silver-tipped staff used by higher-ranking cabildo members as symbol of office

**batelón** (<Brazilian Portuguese **batelão**): a larger river craft with a rudder and a covered shelter at the stern

**Baure:** an Arawakan-speaking indigenous group of the Llanos de Mojos, settled at Concepción, San Joaquín, and Carmen

**bebedero:** a precontact Mojo building used for religious purposes including consumption of alcohol

**bolacha** (<Brazilian Portuguese **borracha**): a mass of coagulated raw rubber; also called plancha

**boliviano:** a decimal-based currency adopted in 1864 to replace the old peso of eight reales; divided into 100 centécimos (renamed centavos in 1870)

**cabildo:** colonial term for a Spanish town council, also applied to the native governing councils created to administer Indian towns in Spanish fashion

**cachuela** (<Brazilian Portuguese **cachoeira**): rapids or waterfall

**cacique:** the head of an indigenous cabildo

**caja:** a drum made from a tree trunk

**camba** (<Guaraní): originally a derogatory term for a Christianized Indian peasant in the Oriente, now also any person from Santa Cruz or lowland regions settled by Cruceños

**camijeta** (<Spanish **camiseta**): man's cotton tunic

**Canichana:** an indigenous group of the Llanos de Mojos, settled in the town of San Pedro; their language is an isolate

**cantón:** administrative district of a former mission town, headed by a corregidor; a subdivision of a province, which was in turn a subdivision of a department

**carayana** (<Guaraní **karai** + Arawakan plural **–ana**): Mojo term for a white person of Spanish descent

**Caripuna:** a Panoan-speaking indigenous group of the northern rainforest; lived independently around the Madeira-Mamoré cachuelas

**catastro:** a register of rural properties

**Cayubaba:** an indigenous group of the Llanos de Mojos, settled at Exaltación; their language is an isolate

**chaco:** a slash-and-burn agricultural plot used in shifting cultivation

**Chacobo:** a Panoan-speaking indigenous group of the Llanos de Mojos; lived independently northwest of Exaltación

**chicha:** a fermented alcoholic beverage, usually made from corn

**Chimane:** a Mosetén-Chonan-speaking indigenous group of the Llanos de Mojos; lived independently west of San Ignacio

**Chiquito:** an indigenous people in eastern Santa Cruz department, commonly called Chiquitano in recent times, settled at Jesuit missions in the colonial period

**Chiquitos:** a province in eastern Santa Cruz department

**chivé** (<Brazilian Portuguese **xibe**): manioc flour mixed with water

**choquigua** (<Itonama **chukiva**): Itonama term for a spirit

**choquigüero:** Spanish/Itonama term for an Itonama shaman

**colegio:** conventual buildings attached to the mission church, included priests' residence and workshops; often became seat of civil authorities

**colla:** a person from the highlands of Bolivia

**colono:** a tenant farmer

**comocoe:** a precontact Mojo shaman who had survived a jaguar attack and acquired the ability to communicate with jaguar spirits

**corregidor:** a government official appointed by a departmental prefect to govern a cantón

**criado:** a household servant raised by a wealthy family almost as its own child, but with the expectation of subservience and domestic service; could be orphans, unwanted children, or captured "bárbaro" children

**Cruceño:** someone from Santa Cruz, or a descendant of someone from Santa Cruz

**cuartel:** a long, rectangular, multifamily house

**curiche** (<Chané, an Arawakan language from Santa Cruz): a depression or meander cutoff filled with water year-round

**enganchador:** a labor recruiter

**enganche:** labor recruitment through cabildo officials or through debt created by an advance

**Esse Ejja:** a Tacanan-speaking indigenous group that lived independently on the Madidi and Madre de Dios rivers; also known as Guarayo

**establecimiento:** a sugar plantation in Beni

**estancia:** an outlying cattle-raising settlement belonging to an ex-mission of Mojos; later, a privately owned ranch

**estrada** (<Brazilian Portuguese): a loop-shaped trail from a base camp though the forest linking several isolated rubber trees

**familia:** the native elite of the Mojos missions, comprised of cabildo officers, religious assistants, and artisans

**fiestas patrias:** the annual celebration of Bolivian independence on August 6

**fiscal:** attorney general of a department or of the nation; also, the title of an indigenous cabildo officer charged with keeping order

**forastero:** an immigrant, someone not native to the district in which they reside

**garitea** (<Brazilian Portuguese **igarité**): a river craft with a rudder and a covered shelter at the stern, smaller than a batelón

**Guarayo:** a Tupí-Guaraní-speaking indigenous group settled into Franciscan missions midway between the Mojos and Chiquitos, near the boundary between Beni and Santa Cruz; also, an alternate name for the unrelated Esse Ejja people

**impuesto de prestación vial:** a tax (created in 1905) to support road work or canal building, which could be paid in either money or through actual labor on those projects

**isla:** a patch of high ground that remains above floodwaters

**Itonama:** an indigenous group of the Llanos de Mojos settled at Magdalena and Huacaraje, also called Machoto; their language is an isolate

**league:** a variable unit of measurement originally based on how far a person could walk or travel on horseback in an hour; one version was equal to 5,000 varas or to 4.2 kilometers (2.6 miles), while others measured 5.6 kilometers (3.5 miles) and 5.9 kilometers (3.7 miles)

**lengüeta:** the cattle disease anthrax

**macana:** a type of ordinary cotton textile

**machetero:** a ritual dance performed with a machetelike wooden implement, or the person who performs it; the dance is also called chiriperono and tontochis

**matrícula:** an official registry of people

**matrícula de mozos:** an official registry of laborers in Beni which specified their patrones and terms of service, sanctioned during the rule of the Liberal Party (1899–1920)

**mesa común:** communal food supplies available to all in the ex-missions

**milréis:** monetary unit of Brazil; originally equal to twenty-seven British pence, its value declined continuously into the early twentieth century

**Mojo:** an Arawakan-speaking indigenous group of the Llanos de Mojos, settled in the towns of Loreto, Trinidad, San Ignacio, and San Javier; commonly called Mojeño in recent times

**More:** an indigenous group that lived independently between the Mamoré and Iténez/Guapore rivers, also called Iténez

**motacú** (<Chiquito): a type of palm; its leaves are used for thatching roofs

**Movima:** an indigenous group of the Llanos de Mojos, settled at Santa Ana; their language is an isolate

**mozo:** a servant, used for indigenous or mestizo laborers on ranches and sugar plantations, sometimes called a colono (tenant farmer)

**Mura:** an indigenous group of the Madeira River

**omiso:** a person who failed to provide proof of military registration after 1904, subject to forced labor if caught

**Oriente:** the eastern two-thirds of Bolivia, comprised of the Departments of Santa Cruz, Beni, and Pando, along with parts of several highland-centered departments

**originario:** an indigenous person native to a district

**Pacahuara:** a Panoan-speaking indigenous group of the northern rainforest; lived independently between the middle Abuná and lower Mamoré/Guapore rivers

**padrón:** a census of tributaries

**parcialidad:** at Spanish contact, a small ethnic group; in the mission period, a group within the pueblo order, supposedly descended from such small ethnic groups; after the liberal reforms, any division of society, whether descended from pueblo parcialidades or familia industries

**Parintintin:** a Tupí-Guaraní-speaking indigenous group of the Madeira River

**patrón:** boss; could be a merchant leading a trade expedition, a rancher, a sugar planter, or someone using workers to extract rubber

**Pauserna:** a Tupí-Guaraní-speaking indigenous group that lived independently on the upper Iténez River, also called Guarasug'wé

**peso:** monetary unit of the Spanish colonies and early republican Bolivia, divided into eight reales; officially replaced by the boliviano in 1864, but documents continue to refer to pesos for several decades

**peso feble:** a debased silver coin used in early republican Bolivia

**prefect:** head of a department, appointed by the national government

**pueblo:** the commoners of the Mojos missions, comprised of workers engaged in agriculture, cattle raising, and river transport

**quintal:** a unit of weight equal to 4 arrobas, or about 45.36 kilograms (100 pounds)

**reservado:** an indigenous male exempt from tribute for being fifty years old or older, or for performing services for the church

**siringuero** (<Brazilian Portuguese **seringueiro**): a rubber tapper

**Sirionó:** a Tupí-Guaraní-speaking hunter-gatherer people who lived east of the Mamoré

**subcorregidor:** one title for a government official heading a vice cantón

**surazo:** a cold south wind that reaches the Llanos de Mojos from Argentina

**tabique:** wattle and daub construction

**Tacana:** a Tacanan-speaking indigenous group of the northern rainforest; lived independently west of the middle Beni River

**tacuara:** a type of cane or bamboo used in construction

**temporalidades:** a system of production and redistribution employed in the Mojos missions until the mid-nineteenth century, in which Indians produced cotton cloth, cacao, and other local products in exchange for salt, iron, and other exotic products

**tiarauqui:** general term for a precontact Mojo shaman

**tierras baldías:** lands which the national government considered unoccupied and hence national property that could be granted to petitioners; included all the lands used by independent "bárbaros" as well as some poorly defined lands of ex-mission Indians

**tipoy** (<Guaraní): woman's cotton tunic

**torito:** a ritual dance where one participant dresses as a bull, descended from an earlier takiríkare dance

**Toromona:** a Tacanan-speaking indigenous group of the northern rainforest; lived independently between the Beni and Madre de Dios rivers

**tutuma:** a gourd or gourd vessel

**vara:** black staff used by lower-ranking cabildo officials as symbol of office; also, a unit of measurement corresponding to a yard

**vecino:** colonial term for a town resident or householder, implying Spanish ancestry and membership in a town's elite; partly superseded by the term *citizen* after the liberal reforms

**vice cantón:** a subdivision of a cantón, headed by a subcorregidor or alcalde de campo

**visita:** a government inspection

**visitador:** a government inspector

**yangada** (<Brazilian Portuguese **jangada**): a raft made by joining two batelones together, used to transport cattle

**yomomo:** a body of water completely covered with floating plants

**Yuracare:** an indigenous group of the Andean foothills in Cochabamba department; their language is an isolate

# *Bibliography*

## Unpublished Primary Sources

Archivo y Biblioteca Histórico del Arzobispado de Santa Cruz (ABHASC)
Archivo Nacional de Bolivia (ANB)
  Ministerio del Interior (MI)
  Revisitas (RV)
  Tribunal Nacional de Cuentas (TNC)
Biblioteca y Archivo Histórico Universidad Autónoma Gabriel Rene Moreno (BAHUAGRM)
Centro Bibliográfico Documental Histórico Chuquisaca (CBDHCh)
Serie correspondencia de la prefectura (SCP)
Museu de Arte Moderna, Rio de Janeiro (MAM-Rio)

## Published Primary Sources


Agassiz, Louis, and Elizabeth Cabot Cary. A *Journey in Brazil*. Boston: Tickner and Fields, 1868.

Altamirano, Diego Francisco. *Historia de la Misión de los Mójos*. La Paz: Instituto Boliviano de Cultura, 1979.

*Anexos a la memoria presentada en la legislatura de 1907*. La Paz: Tip. Artística de Castillo y C$^a$, 1908.

*Anuario de leyes, decretos, resoluciones y órdenes supremas, año de 1905 (edición oficial)*. La Paz: Tipografía Artística de Castillo y C$^a$., 1906.

*Anuario de leyes y supremas disposiciones de 1896 (compilado por Claudio Q. Barrios), edición oficial*. La Paz: Imprenta de "El Comercio," 1898.

*Anuario de leyes y supremas disposiciones de 1897*. n.d.

Arauz, Ygnacio. *Nueva via fluvial de Bolivia*. Manaos: Typographia de "Amazonas" de A. de C. Mendes, 1868.

Armentia, Fray Nicolás. *Diario de sus viages a las tribus comprendidas entre el Beni y Madre de Dios y en el Arroyo de Ivon en los años de 1881 y 1882*. La Paz: "Tipografía Religiosa," 1883.

Arteche, Gumercindo Gómez de. *JHS misión de los P.P. Astrain, Manzanedo y Arteche, 1888*. Jorge Cortés Rodríguez, ed. Trinidad: CIDDEBENI, 1989.

Barros Cavalcanti de Albuquerque Lacerda, Adolfo de. *Relatorio apresentado a'assemblea legislativa da provincia do Amazonas na sessão ordinario do 1° de outubro de 1864*. Pernambuco: Typographia de Manoel Figueiroa de Faria and Filho, 1864.

———. *Relatorio com que o illustrissimo e excellentissimo senhor Dr. Adolpho de Barros Cavalcanti de A. Lacerda entregou a administração da provincia do Amazonas ao illm. e exm. sr. Tenente Coronel Innocencio Eustaquio Ferreira de Araujo*. Recife: Typ. do Jornal do Recife, 1865.

Benjamin, Feliciano Antonio. *Relatorio apresentado ao exm. sr. Dr. Presidente da Provincia Domingos Monteiro Peixoto, pelo Engenheiro Bacharel Feliciano Antonio Benjamin, membro da Commissão de fiscalisação da estrada de ferro do Madeira e Mamoré e de medição e demarcação de terras no Rio Madeira, actualmente encarregado da mesma commissão*. Manaus: Typographia do Commercio do Amasonas, 1874.

Boccolini, Rosa, and Rosario Jiménez. *Estadísticas de contribución indígena en Bolivia, 1770–1902*. Vol. 2. Lima: Biblioteca Andina, 1979.

*Boletín del Ministerio de Relaciones Esteriores y Colonización*. 15 febrero 1887.

Bowman, Isaiah. "Geographical Aspects of the New Madeira-Mamoré Railroad." *Bulletin of the American Geographical Society* 45, no. 4 (1913): 275–81.

Brown, Charles Barrington, and William Lidstone. *Fifteen Thousand Miles on the Amazon and its Tributaries*. London: Stanford, 1878.

Cabrera, Ladislao. *Navegación fluvial de Trinidad a Villa Bella. Diario de viaje por las cachuelas del Mamoré en el Departamento del Beni*. Santa Cruz: La Estrella del Oriente, 1889.

Cámara de Comercio e Industria Trinidad. *Informe de los delegados del Departamento del Beni, que asistieron al primer congreso nacional de agricultura y ganadería*. La Paz: Editorial "America," 1931.

Cardús, Fray José. *Las misiones franciscanas entre los infieles de Bolivia*. Barcelona: Librería de la Inmaculada Concepción, 1886.

Carneiro da Cunha, Manoel Clementino. *Relatorio apresentado á Assemblea Legislativa da provincia do Amazonas pelo Exm° Senr. Dr. Manoel Clementino Carneiro da Cunha, presidente da mesma provincia, na sessão ordinaria de 3 de maio de 1862*. Pará: Typographia de Frederico Carlos Rhossard, 1862.

Church, George Earl. *Aborigines of South America*. London: Chapman and Hall, 1912.

Coimbra, Juan B. *Siringa*. La Paz: Librería Editorial America, 1989.

Comissão de Estudos da Estrada de Ferro do Madeira e Mamoré. *Do Rio de Janeiro ao Amazonas e Alto Madeira; itinerario e trabalhos da comissão de estudos da Estrada de Ferro do Madeira e Mamoré: impressões de viagem por um dos membros da mesma comissão*. Rio de Janeiro: Typ. a Vap. de Soares and Niemeyer, 1885.

Craig, Neville B. *Recollections of an Ill-Fated Expedition to the Headwaters of the Madeira River in Brazil*. Philadelphia: J. B. Lippincott, 1907.

Ducci, Zacarias. *Diario de la visita a todas las misiones existes en la República de Bolivia—América Meridional*. Asís: Tipografía de la Porciúncula, 1895.

Eder, Francisco J. *Breve descripción de las reducciones de Mojos*. Cochabamba: Historia Boliviana, 1985.

Fonseca, João Severino da. *Voyage autour du Brésil*. Rio de Janeiro: Librairie A. Lavignasse Filho, 1899.

Furtado, Francisco José. *Relatorio que a'assemblea legislativa provincial do Amazonas apresentou na abertura da sessão ordinaria em o dia 3 de maio de 1859*. Manaus: Typographia de Francisco José da Silva Ramos, 1859.

Gibbon, Lardner. *Exploration of the Valley of the Amazon, Made under Direction of the Navy Department, by Wm. Lewis Herndon and Lardner Gibbon, Lieutenants United States Navy*. Part II. Washington, DC: A.O.P. Nicholson, 1854.

Grubb, Kenneth. *Amazon and Andes*. New York: L. Mac Veagh, Dial Press, 1930.

Gutiérrez, José Manuel. *Colección oficial del año de 1858, que contiene todas las leyes, decretos, órdenes supremas y resoluciones, que se han dado, desde el 9 de diciembre de 1857 hasta el 31 de diciembre de 1858, por el excelentísimo gobierno provisorio de la república*. Cochabamba: Tipografía de Quevedo, 1859.

Jacy Monteiro, Domingos. *Relatorio apresentado ao exm^o^ Sr. Dr. Agesiláo Pereira da Silva, presidente da provincia do Amazonas pelo Dr. Domingos Jacy Monteiro, depois de ter entregue a admimistração [sic] da provincia em 26 de maio de 1877*. Manaus: Typ. do "Amazonas" de José Carneiro dos Santos, 1878.

Keller, José y Francisco. *Exploración del Río Madera en la parte comprendida entre la Cachuela de San Antonio y la embocadura del Mamorè, por los injenieros brasileros José y Francisco Keller*. La Paz: Imprenta de la Union Americana—Por Cèsar Sevilla, 1870.

Keller-Leuzinger, Franz. *The Amazon and Madeira Rivers*. Philadelphia: J. B. Lippincott, 1875.

Lacerda e Almeida, Francisco José de. "Memoria a respeito dos Rios Baures, Branco, da Conceição, de S. Joaquim, Itonamas e Maxupo; e das tres missões da Magdalena, da Conceição e de S. Joaquim." *Revista do Instituto Histórico e Geográfico Brasileiro* 12 (1849): 106–19.

Lens, Benjamín. *El Departamento del Beni desde fines del 70 hasta Febrero del 71*. La Paz: Imprenta de la Union Americana—de César Sevilla, 1872.

Maccheti, Fray Jesualdo. *Diario del viaje fluvial del Padre Fray Jesualdo Maccheti, misionero del Colejio de La Paz, desde San Buenaventura y Réyes hasta el Atlántico en 1869*. La Paz: Imprenta de "El Siglo Industrial," n.d.

Maracajú, Barão de. *Falla com que abrio no dia 25 de agosto de 1878 a 1ª sessão da 14ª legislatura da Assembléa Legislativa Provincial do Amazonas o exmº sr. Barão de Maracajú, presidente da provincia*. Manaus: Typ. do Amazonas, 1878.

———. *Falla com que o exmº sr. Barão de Maracajú, presidente da provincia do Amazonas, no dia 29 de março de 1879 abriu a 2ª sessão da 14ª legislatura da Assembléa Legislativa Provincial*. Manaus: Typ. do Amazonas, 1879.

Martins de Menezes, Aprigio. *Relatorio apresentado ao exm. sr. Dr. Presidente da Provincia Domingos Monteiro Peixoto pelo Dr. Aprigio Martins de Menezes, encarregado da Enfermaria dos Variolosos*. Manaus: Typographia do Commercio do Amasonas, 1874.

Martins da Silva Coutinho, Luiz. *Relatorio apresentado ao Illm. e Exm. Sr. Presidente da provincial do Amasonas, Dr. José de Miranda da Silva Reis pelo Director das Obras Publicas da mesma provincial, Luiz Martins da Silva Coutinho.* Manaus: Typographia de Gregorio José de Moraes, 1871.

Mathews, Edward Davis. *Up the Amazon and Madeira Rivers through Bolivia and Peru.* London: S. Low, Marston, Searle and Rivington, 1879.

Miranda da Silva Reis, José de. *Relatorio apresentado a Assembléa Legislativa Provincial do Amasonas na primeira sessão da 11ª legislatura no dia 25 de março de 1872 pelo presidente da provincia o Exmº Sr. Dr. José de Miranda da Silva Reis.* Manaus: Typographia de Gregorio José de Moraes, 1872.

Monteiro Peixoto, Domingos. *Falla diririgida [sic] á Assembléa Legislativa Provincial do Amasonas na segunda sessão da 11ª Legislatura em 25 de março de 1873 pelo presidente da provincia, bacharel Domingos Monteiro Peixoto.* Manaus: Typ. do Commercio do Amazonas de Gregorio Joze de Moraes, 1873.

———. *Relatorio com que o Exm. Sr. Dr. Domingos Monteiro Peixoto entregou a Administração da Provincia ao Exm. Sr. 1º vice-presidente capitão de mar e guerra Nuno Alves Pereira de Mello Cardoso em 16 de março do 1875.* Manaus: Typ. do Commercio do Amasonas, 1875.

Nordenskiöld, Erland. *Indianer und Weisse in Nordostbolivien.* Stuttgart: Strecker und Schröder, 1923.

Oficina Nacional de Inmigración, Estadística y Propaganda Geográfica. *Censo general de la población de la República de Bolivia según el empadronamiento de 1º de septiembre de 1900.* 2 vols. La Paz: Taller Tipo-Litográfico de José M. Gamarra, 1902, 1904.

———. *Censo general de la población de la República de Bolivia según el empadronamiento de 1º de septiembre de 1900.* 2 vols. Reprint, La Paz, Cochabamba: Editorial Canelas S.A., 1973.

———. *Sinopsis estadística y geográfica de la República Boliviana.* Tomo II. La Paz: Tall. Tip.-Lit. de J.M. Gamarra, 1903.

Orbigny, Alcides d'. *Descripción geográfica, histórica y estadística de Bolivia.* Tomo I. La Paz: Instituto Cultural Anglo-Boliviano, 1946.

———. *El hombre americano.* Buenos Aires: Editorial Futuro, 1944.

———. *Viaje a la América meridional . . . realizado de 1826 a 1833.* Buenos Aires: Editorial Futuro, 1945.

Orton, James. *The Andes and the Amazon.* New York: Harper and Brothers, 1870.

Pesciotti, Bernardino J. *Informe que presenta al Supremo Gobierno el R.P. Prefecto Fr. Bernardino J. Pesciotti sobre el movimiento de las misiones de su cargo.* Santa Cruz: Tipografía Industrial, 1914.

Pinkas, Júlio. *Relatorio apresentado a S. Ex. o Sr. Conselheiro João Ferreira de Moura, Ministro e Secretario de Estado dos Negocios da Agricultura, Commercio e Obras Publicas.* Rio de Janeiro: Imprensa Nacional, 1885.

René Moreno, Gabriel. *Catálogo del Archivo de Mojos y Chiquitos.* La Paz: Libreria Editorial "Juventud," 1974.

República de Bolivia. *Anuario de leyes, decretos y resoluciones supremas de 1912,* compilado por Mariano Rivas L. Tomo I. La Paz: Talleres Gráficos "La Prensa," 1913.

Román, Federico. *Ferrocarril Cochabamba-Santa Cruz. Razones en pro del trazo del norte.* La Paz: Escuela Tipográfica Salesiana, 1922.

Rusby, Henry Hurd. *Jungle Memories*. New York: Whittlesey House, 1933.
Salinas, Manuel Macedonio. *Navegación de los ríos de Bolivia confluentes del Madera y Amazonas y colonización*. Cochabamba: Imprenta de Gutierrez, 1871.
Schurz, W. L. "The Distribution of Population in the Amazon Valley." *Geographical Review* 15, no. 2 (April 1925): 206–25.
Souza, Cônego Francisco Bernardino de. *Lembranças e curiosidades do valle do Amazonas*. Pará: Typografía do Futuro, 1873.
Suárez, Daniel. *Manifiesto del ex-prefecto del Beni ante la opinión pública*. Trinidad: Imprenta Suarez y Hnos., 1887.
*Sublevación de los Indios Chiriguanos en las provincias de Azero y Cordillera, pertenecientes a los Departamentos de Sucre y Santa-Cruz de la República de Bolivia*. Potosí: Imp. de "El Porvenir," 1892.
Tavares Bastos, Aureliano Cândido. *O vale do Amazonas*. São Paulo: Companhia Editora Nacional, 1975.
Wilkens de Mattos, João. *Relatorio com que o Exm° Sr. Presidente da provincial do Amazonas, Teniente-Coronel João Wilkens de Mattos, abrio a Assambléa Legislativa Provincial no dia 4 de abril de 1869*. Manaus: Typographia do Amazonas de Antonio da Cunha Mendes, 1869.
———. *Relatorio lido pelo Exm° Sr. Presidente da Provincia do Amazonas, tenente-coronel João Wilkens de Mattos, na sessão d'abertura da Assembléa Legislativa Provincial á 25 de março de 1870*. Manaus: Typographia do Amazonas de Antonio da Cunha Mendes, 1870.

## Periodicals

*Archivo de la Comisaria Franciscana de Bolivia*
*Crónica Guaraya*

## Newspapers

*El Correo del Beni* (Trinidad)
*Los Debates* (Sucre)
*La Democracia* (Trinidad)
*El Eco del Beni* (Trinidad)
*La Estrella del Oriente* (Santa Cruz)
*La Voz del Pueblo* (Trinidad)

## Secondary Sources

Aalto, Rolf, et al. "Episodic Sediment Accumulation on Amazonian Flood Plains Influenced by El Niño/Southern Oscillation." *Nature* 425 (2 Oct. 2003): 493–97.
Albó, Xavier, et al. *Para comprender las culturas rurales en Bolivia*. La Paz: MEC-CIPCA-UNICEF, 1989.
Alonso, Ana María. *Thread of Blood*. Tucson: University of Arizona Press, 1995.

Anderson, Gary C. *The Indian Southwest, 1580–1830*. Norman: University of Oklahoma Press, 1999.

Apter, Andrew. *Beyond Words*. Chicago: University of Chicago Press, 2007.

Armani, Alberto. *Ciudad de Dios y Ciudad del Sol*. México: Fondo de Cultura Económica, 1982.

Artz, Frederick B. *Reaction and Revolution, 1814–1832*. New York: Harper and Row, 1934.

Banzer Toro de Añez, Emma. *Monografía de Exaltación de la Santa Cruz*. Santa Cruz: Impresiones Offset Color "Oriente," 2004.

Barham, Bradford L., and Oliver T. Coomes. *Prosperity's Promise*. Boulder, CO: Westview Press, 1996.

Barragán, Rossana. "Ciudadanía y elecciones, convenciones y debates." In Rossana Barragán and José Luis Roca, *Regiones y poder constituyente en Bolivia*. La Paz: Informe sobre el Desarrollo Humano en Bolivia, 2005.

Becerra Casanovas, Rogers. *Reliquias de Moxos*. La Paz: Empresa Editora "Proinsa," 1978.

Beezley, William H., et al., eds. *Rituals of Rule, Rituals of Resistance*. Wilmington, DE: Scholarly Resources, 1994.

Billeter, Erika. *Canto a la realidad*. Madrid: Casa de América, 1993.

Block, David. *Mission Culture on the Upper Amazon*. Lincoln: University of Nebraska Press, 1994.

Brooks, James F. *Captives and Cousins*. Chapel Hill: University of North Carolina Press, 2002.

Brown, Michael F. "Beyond Resistance: Comparative Study of Utopian Renewal in Amazonia." In Anna Roosevelt, ed. *Amazonian Indians from Prehistory to the Present*. Tucson: University of Arizona Press, 1994.

———. "On Resisting Resistance." *American Anthropologist* 98, no. 4 (Dec. 1996): 729–35.

Buecher, Hans C., and Judith-Maria Buecher. *The Bolivian Aymara*. New York: Holt, Rinehart, and Winston, 1971.

Bullock, Alan, and Maurice Shock, eds. *The Liberal Tradition*. London: Adam and Charles Black, 1956.

Burns, E. Bradford. *A History of Brazil*. New York: Columbia University Press, 1980.

———. *The Poverty of Progress*. Berkeley: University of California Press, 1980.

Buve, Raymond. "Political Patronage and Politics at the Village Level in Central Mexico." *Bulletin of Latin American Research* 11, no. 1 (January 1992): 1–28.

Callaú Barbery, Ignacio. "La Guayochería." *Revista de la Universidad Autónoma Gabriel René Moreno* (Santa Cruz) 3, no. 8 (Dic. 1950): 177–81.

Cañeque, Alejandro. *The King's Living Image*. New York: Routledge, 2004.

Carrasco, Manuel. *José Ballivián, 1805–1852*. Buenos Aires: Machette, 1960.

Carvalho Urey, Antonio. *Visión del Beni*. Cochabamba: Editorial Serrano, 1978.

———. *Beni: Ensayo de interpretación histórica*. La Paz: Imprenta y Libreria "Renovación," 1983.

Centeno, Ricardo, and Patricia Fernández O. *Imágenes del auge de la goma*. La Paz: La Papelera, 1998.

Central de Pueblos Indígenas del Beni, Centro de Información y Documentación para el Desarrollo del Beni. *Hacia una propuesta indígena de descentralización del estado*. La Paz: PROADE: ILDIS, 1995.

Coates, Austin. *The Commerce in Rubber.* Singapore: Oxford University Press, 1987.

Cohn, Norman. *The Pursuit of the Millennium.* New York: Oxford University Press, 1970.

Condori Chura, Leandro, and Esteban Ticona Alejo. *El escribano de los caciques apoderados.* La Paz: Hisbol/THOA, 1992.

Crapanzano, Vincent. "Spirit Possession." In Mircea Eliade, ed., *The Encyclopedia of Religion.* New York: MacMillan, 1987.

Dean, Warren. *Brazil and the Struggle for Rubber.* Cambridge: Cambridge University Press, 1987.

Denevan, William M. *The Aboriginal Cultural Geography of the Llanos de Mojos of Bolivia.* Berkeley: University of California Press, 1966.

Díez Astete, Alvaro, and David Murillo. *Pueblos indígenas de tierras bajas.* La Paz: Ministerio de Desarrollo Sostenible y Planificación, Viceministerio de Asuntos Indígenas y Pueblos Originarios, Programa Indígena-PNUD, 1998.

Ducey, Michael T. "Liberal Theory and Peasant Practice." In Robert H. Jackson, ed. *Liberals, the Church, and Indian Peasants: Corporate Lands and the Challenge of Reform in Nineteenth-Century Spanish America.* Albuquerque: University of New Mexico Press, 1997.

Eliade, Mircea. *Shamanism.* Princeton: Princeton University Press, 1964.

——, ed. *The Encyclopedia of Religion.* New York: MacMillan, 1987.

Fifer, J. Valerie. "The Empire Builders." *Journal of Latin American Studies* 2, no. 2 (1970): 113–46.

Flores Moncayo, José. *Legislación boliviana del indio.* La Paz: Ministerio de Asuntos Campesinos, Departamento de Publicaciones del Instituto Indigenista Boliviano, 1953.

Gamarra Tellez, María del Pilar. "Haciendas y peones en el régimen hacendatario gomero boliviano." Tesis de Licenciatura, Universidad Mayor de San Andrés, La Paz, 1992.

——. "La participación estatal en la industria de la goma elástica." *DATA, Revista del Instituto de Estudios Andinos y Amazónicos* 4 (1993): 23–75.

García, María Elena. *Making Indigenous Citizens.* Stanford: Stanford University Press, 2005.

Gotkowitz, Laura. "Within the Boundaries of Equality." PhD diss., University of Chicago, 1998.

——. *A Revolution for Our Rights.* Durham, NC: Duke University Press, 2007.

Gow, Peter. "River People." In Nicholas Thomas and Caroline Humphrey, eds. *Shamanism, History, and the State.* Ann Arbor: University of Michigan Press, 1994.

Gray, John. *Liberalism.* Minneapolis: University of Minnesota Press, 1986.

Greever, Janet Groff. *José Ballivián y el oriente boliviano.* La Paz: Empresa Editora Siglo, 1987.

Grubb, Kenneth G. *Amazon and Andes.* New York: Dial Press, 1930.

Guy, Donna J., and Thomas E. Sheridan, eds. *Contested Ground.* Tucson: University of Arizona Press, 1998.

Heath, Dwight B. "Camba." PhD diss., Yale University, 1959.

Hemming, John. *Amazon Frontier.* London: Macmillan, 1987.

Hobsbawm, Eric J. *Primitive Rebels.* New York: W. W. Norton, 1959.

——. *Industry and Empire.* Harmondsworth: Penguin Books, 1968.

Hollander, Jocelyn A., and Rachel L. Einwohner. "Conceptualizing Resistance." *Sociological Forum* 19, no. 4 (Dec. 2004): 533–54.

Holmberg, Allan R. *Nomads of the Long Bow*. Garden City, NY: Natural History Press, 1969.

Hornborg, Alf. "Ethnogenesis, Regional Integration, and Ecology in Prehistoric Amazonia." *Current Anthropology* 46, no. 4 (Aug.–Oct. 2005): 589–602.

Hugh-Jones, Stephen. "Shamans, Prophets, Priests, and Pastors." In Nicholas Thomas and Caroline Humphrey, eds. *Shamanism, History, and the State*. Ann Arbor: University of Michigan Press, 1994.

Hurtado, Albert L. *Indian Survival on the California Frontier*. New Haven: Yale University Press, 1988.

Ingold, Tim. *Hunters, Pastoralists, and Ranchers*. Cambridge: Cambridge University Press, 1980.

Jackson, Robert H. *Regional Markets and Agrarian Transformation in Bolivia*. Albuquerque: University of New Mexico Press, 1994.

——, ed. *Liberals, the Church, and Indian Peasants*. Albuquerque: University of New Mexico Press, 1997.

——, ed. *New Views of Borderlands History*. Albuquerque: University of New Mexico Press, 1998.

Jackson, Robert H., and Edward Castillo. *Indians, Franciscans, and Spanish Colonization*. Albuquerque: University of New Mexico Press, 1995.

Jamieson, Ross W. *Domestic Architecture and Power*. New York: Kluwer Academic/Plenum, 2000.

Jones, James C. "Conflict between Whites and Indians on the Llanos de Moxos, Beni Department." PhD diss., University of Florida, 1980.

Jordá, Enrique, et al., eds. *Historia del pueblo Mojo*. La Paz: Talleres Gráficos de Editorial Popular, 1989.

——. "La Semana Santa y la cruz en los pueblos Mojos del Beni (Bolivia)." In Frans Damen and Esteban Judd Zanon, eds., *Cristo crucificado en los pueblos de América Latina*. Cusco-Sicuani: Ed. Instituto de Pastoral Andina; Quito: Ed. ABYA-YALA, 1992.

Jordan, Terry. *North American Cattle Ranching Frontiers*. Albuquerque: University of New Mexico Press, 1993.

Kicza, John E., ed. *The Indian in Latin American History*. Wilmington, DE: Scholarly Resources, 1993.

Klein, Herbert S. *Bolivia*. New York: Oxford University Press, 1992.

Langer, Erick. *Economic Change and Rural Resistance in Southern Bolivia, 1880–1930*. Stanford: Stanford University Press, 1989.

——. *Expecting Pears from an Elm Tree*. Durham, NC: Duke University Press, 2009.

Langer, Erick, and Robert H. Jackson. "Colonial and Republican Missions Compared." *Comparative Studies in Society and History* 30, no. 2 (Apr. 1988): 289–303.

Langer, Erick, and Robert H. Jackson, "Liberalism and the Land Question in Bolivia, 1825–1920." In Robert H. Jackson, ed. *Liberals, the Church, and Indian Peasants*. Albuquerque: University of New Mexico Press, 1997.

Larson, Brooke. *Trials of Nation Making*. Cambridge: Cambridge University Press, 2004.

Legrand, Catherine C. "Informal Resistance on a Dominican Sugar Plantation during the Trujillo Dictatorship." *Hispanic American Historical Review* 75, no. 4 (Nov. 1995): 555–96.
Lehm Ardaya, Zulema. "Proyecto: 'Diagnostico de la situación actual de los indígenas de Trinidad y areas cercanas.'" Trinidad: CIDDEBENI, 1987.
——. *Milenarismo y moviemientos sociales en la Amazonia boliviana*. Santa Cruz: CIDDEBENI, APCOB, OXFAM América, 1999.
Lehm Ardaya, Zulema, Arnaldo Lijerón Casanovas, and Lorenzo Vare Chávez. *Diagnóstico socio-económico de los indígenas mojeños en la ciudad de Trinidad*. Trinidad: CIDDEBENI, 1990.
Lijerón Casanovas, Arnaldo. *Facetas de la cultura beniana*. Trinidad: Talleres de la Oficina Técnica de Poligrafiados, 1982.
Limpias Saucedo, Manuel. *Los gobernadores de Mojos*. La Paz: Escuela Tipográfica Salesiana, 1942.
Mendizabal, Fray Santiago. *Vicariato Apostólico del Beni*. La Paz: Imp. "Renacimiento," 1932.
Métraux, Alfred. "The Social Organization and Religion of the Mojo and Manasi." *Primitive Man* 16, nos. 1 and 2 (January and April, 1943): 1–30.
Monje Roca, Raul. *El río Mamoré*. La Paz: Instituto Boliviano de Cultura, 1977.
Morse, Birgit Faber. "The Salt River Site, St. Croix, at the Time of the Encounter." In Samuel M. Wilson, ed. *The Indigenous People of the Caribbean*. Gainesville: University Press of Florida, 1997.
Muller, Jean-Claude. "La pierre qui parle ou le savoir ventriloque." *Anthropologie et sociétés* 17, no. 3 (1993): 93–101.
Muratorio, Blanca. *The Life and Times of Grandfather Alonso*. New Brunswick, NJ: Rutgers University Press, 1991.
Paola, Chris. "Floods of Record." *Nature* 425 (2 Oct. 2003): 459.
Paravicini, Luis Carlos. *Diccionario legislativo boliviano*. Tomo II. La Paz: Editorial Agiurre, 1991.
Paredes-Candia, Antonio. *La danza folklórica en Bolivia*. La Paz: Libreria-Editorial "Popular," 1991.
Parejas Moreno, Alcides. "Las misiones jesuíticas de Chiquitos." In *Los bolivianos en el tiempo. Cuadernos de historia*. Sucre: Universidad Andina Simón Bolívar, Instituto de Estudios Andinos y Amazónicos, 1993.
Pinto Parada, Rodolfo. *Rumbo al Beni*. Cochabamba: Editorial Serrano, 1978.
——. *Pueblo de leyenda*. Trinidad: Tiempo del Bolivia, 1987.
Poole, Deborah, ed. *Unruly Order*. Boulder, CO: Westview Press, 1994.
Radding Murrieta, Cynthia. *Wandering Peoples*. Durham, NC: Duke University Press, 1997.
Rasnake, Roger N. *Domination and Cultural Resistance*. Durham, NC: Duke University Press, 1988.
Rausch, Jane M. A *Tropical Plains Frontier*. Albuquerque: University of New Mexico Press, 1984.
——. *The Llanos Frontier in Colombian History, 1830–1930*. Albuquerque: University of New Mexico Press, 1993.
Reed, Nelson. *The Caste War of Yucatan*. Stanford: Stanford University Press, 1964.

República de Bolivia, Ministerio de Hacienda. *Características demográficas de la población en Bolivia*. La Paz: 1997.

Reséndez, Andrés. *Changing National Identities at the Frontier*. Cambridge: Cambridge University Press, 2005.

Riester, Jürgen. *En busca de la Loma Santa*. La Paz: Editorial Los Amigos del Libro, 1976.

Rivera Cusicanqui, Silvia. "'Pedimos la revisión de límites.' Un episodio de incomunicación de castas en el movimiento de caciques-apoderados en los Andes Bolivianos, 1919–1921." In Segundo Moreno Y. and Frank Salomon, compiladores. *Reproducción y transformación de las sociedades andinas, siglos XVI–XX*. Tomo II. Quito: Ediciones ABYA-YALA, MLAL, 1991.

Roca, José Luis. "Mojos en los albores de la Independencia boliviana (1810–1811)." *Historia y Cultura* 21 (1992): 187–244.

Saignes, Thierry. *Ava y karai*. La Paz: Hisbol, 1990.

Sanabria Fernández, Hernando. *En busca de Eldorado*. Santa Cruz: Universidad Gabriel René Moreno, Buenos Aires: Imprenta López, 1958.

Santamaría, Daniel J. "La economía de las misiones de Moxos y Chiquitos (1675–1810)." *Ibero-Amerikanisches Archiv* 13, no. 2 (1987): 255–95.

Santos Granero, Fernando. *Etnohistoria de la Alta Amazonia*. Quito: Ediciones ABYA-YALA, 1992.

———. *Opresión colonial y resistencia indígena en la Alta Amazonia*. Quito: CEDIME, FLACSO, 1992.

Schapiro, J. S. *Liberalism*. New York: Van Nostrand Reinhold, 1958.

Schoop, Wolfgang. *Ciudades bolivianas*. La Paz, Cochabamba: Editorial Los Amigos del Libro, 1981.

Scott, James C. *Weapons of the Weak*. New Haven: Yale University Press, 1985.

Sheehan, James. *German Liberalism in the Nineteenth Century*. Chicago: University of Chicago Press, 1978.

Siikala, Anna-Leena. "Shamanism: Siberian and Inner Asian Shamanism." In Mircea Eliade, ed. *The Encyclopedia of Religion*. New York: MacMillan, 1987.

Simon, W. M., ed. *French Liberalism 1789–1848*. New York: Wiley, 1972.

Soria Choque, Vitaliano. "Los caciques-apoderados y la lucha por la escuela (1900–1952)." In *Educación indígena: ¿ciudadanía o colonización?* La Paz: Ayuwiyiri, 1992.

Spicer, Edward H. *Cycles of Conquest*. Tucson: University of Arizona Press, 1962.

Stanfield, Michael. *Red Rubber, Bleeding Trees*. Albuquerque: University of New Mexico Press, 1998.

Stern, Steve J., ed. *Resistance, Rebellion, and Consciousness in the Andean Peasant World, 18th to 20th Centuries*. Madison: University of Wisconsin Press, 1987.

Steward, Julian H., ed. *Handbook of South American Indians*. Vol. 2. *The Andean Civilizations*. Washington, DC: US Government Printing Office, 1946.

———. *Handbook of South American Indians*, Vol. 3: *The Tropical Forest Tribes*. Washington, DC: US Government Printing Office, 1948.

Taussig, Michael. *Shamanism, Colonialism, and the Wild Man*. Chicago: University of Chicago Press, 1987.

Thompson, Robert Farris. *Flash of the Spirit*. New York: Vintage Books, 1984.

Thomson, Guy P.C. "Bulwarks of Patriotic Liberalism." *Journal of Latin American Studies* 22 (1990): 31–68.

——. "Agrarian Conflict in the Municipality of Cuetzalán (Sierra de Puebla): The Rise and Fall of 'Pala' Agustín Dieguillo, 1861–1894." *Hispanic American Historical Review* 71, no. 2 (May 1991): 205–58.

——. "Popular Aspects of Liberalism in Mexico, 1848–1888." *Bulletin of Latin American Research* 10, no.3 (1991): 265–92.

Velasco Crespi, Luis. "Fuerzas Armadas." In Josep M. Barnadas, ed. *Diccionario histórico de Bolivia*. Sucre: Grupo de Estudios Históricos, 2002.

"Ventriloquism." In *Collier's Encyclopedia*. Vol. 23. New York: P. F. Collier, 1995.

"Ventriloquism." In *New Encyclopaedia Brittanica*. 15th ed. Vol. 12. Chicago: Encyclopaedia Brittanica, 1997.

Weber, David J. *Bárbaros*. New Haven: Yale University Press, 2005.

Weinstein, Barbara. *The Amazon Rubber Boom, 1850–1920*. Stanford: Stanford University Press, 1983.

Wilson, David J. *Indigenous South Americans of the Past and Present*. Boulder, CO: Westview Press, 1999.

Wolf, Eric R. "Closed Corporate Peasant Communities in Mesoamerica and Java." *Southwestern Journal of Anthropology* 13, no. 1 (1957): 1–18.

# Index

# *About the Author*

**Gary Van Valen** was born in New Jersey. He completed his BA in history at Montclair State College in 1988, his MA in Latin American history at the University of South Carolina in 1995, and his PhD in Latin American history at the University of New Mexico in 2003. After teaching classes for the University of New Mexico, he served as a visiting assistant professor at Roanoke College and as assistant and later associate professor at the University of West Georgia. He teaches courses on Latin American, indigenous, Atlantic, and world history. He has contributed to the *Diccionario histórico de Bolivia*, the *World History Bulletin*, the *SECOLAS Annals*, and the *Anuario de Estudios Bolivianos, Archivísticos y Bibliográficos*. His primary interests are in Latin American indigenous history and the frontiers of Latin America (including borderlands history), with particular emphasis on Bolivia and New Mexico. In both places, his research has centered on the ethnohistory and cultural contact of frontiers. He plans to publish a Spanish translation of *Indigenous Agency in the Amazon* to make these findings more accessible to Bolivians. For his next research project, he will investigate the Pueblos of New Mexico in the period of Mexican rule. This will provide a valuable comparison to his Bolivian research and will further one of his longer-range goals, the publication of a New Mexico history textbook.